Electromagnetism
Nature's Force That Shapes Our Lives

First published by Nottingham University Press

This reissued original edition published 2023 by 5m Books Ltd
www.5mbooks.com

British Library Cataloguing in Publication Data

Electromagnetism: Nature's Force That Shapes Our Lives
L Fagg

ISBN 9781789182781

Typeset by Nottingham University Press, Nottingham

EU GPSR Authorised Representative
LOGOS EUROPE, 9 rue Nicolas Poussin, 17000, LA ROCHELLE, France
E-mail: Contact@logoseurope.eu

Electromagnetism:
Nature's Force That Shapes Our Lives

Lawrence Fagg

Nottingham
University Press

*To Mary in comforting memory
of her courage, joy and compassion*

TABLE OF CONTENTS

ENDORSEMENTS

"Professor Fagg has written an engaging book on the pervasiveness of electromagnetism throughout our natural world. He touches on the extremes of nature and everywhere in between: from the atom to the cosmos; from the very simple to the highly complex; from the practical to the esoteric. In often grand and poetic language, he invites the reader to share in his own personal wonderment at how nature weaves its magic. It was a very enjoyable read."

Prof. AM Nathan
Professor Emeritus, Physics Dept, University of Illinois

"The author presents us with a unique, non-mathematical description of the fundamental ideas and concepts that constitute the intellectual heart of modern physics. Central to his description is the crucial role that is played by electromagnetism, first in the discovery of its governing laws, and then in its application as an experimental tool in probing the fundamental forces of nature and in creating a host of technical applications. His exposition is clear and elegant, adorned with charming personal references and humanizing historical asides. It should be a treasure, accessible to readers at all levels of sophistication."

Calvin M Class
Professor Emeritus of Physics, William Marsh Rice University, Houston, Texas

"This book is an exquisite and touching paean to the many facets of electromagnetism, covering astrophysics and the complexities of the human brain as well as the latest experiments underway at the CERN Large Hadron Collider. After reading this enjoyable book, a layman who is not professionally dependent upon electromagnetism will come away with a new comprehension of what holds our Universe together; and even a theoretical physicist such as myself can read it with profit and pleasure. For myself, I found the descriptions of Chapters 2 and 9 to be superb examples of beautifully-phrased clarity and comprehension.

In brief, this book should be read, and will be enjoyed, by every intelligent person, layman or specialist."

Prof. HM Fried
Professor Emeritus and Research Professor of Physics, Brown University

"This book tells the fascinating story of electromagnetism, its history and its universal presence in the natural world and in the world of technology. Drawing on his vast knowledge and experience as an experimental physicist, Lawrence Fagg has written for the general reader a highly readable, comprehensive account covering all major aspects of the topic. With his deep personal engagement in the subject, he succeeds in communicating to the reader his genuine enthusiasm for electromagnetism."

Erik Balslev

Professor Emeritus, Dept. of Mathematics, Aarhus University, Denmark

"This is an excellent exposition of where modern physics stands and how it got there. Dr. Fagg has had a distinguished career using the scattering of high-energy electrons to study nuclei and he puts his deep understanding of the intellectual pillars upon which physics is built to explain, without using any mathematics, phenomena ranging from the big bang to the silicon diode. The book is very readable and sprinkled with tidbits about the lives of the giants whose contributions have led the development of this great human endeavor. I strongly recommend this book to anyone who wants to have an understanding of the science that governs our basic existence and for anyone contemplating a scientific career there is no better place to start."

Ralph E Segel

Professor Emeritus of Physics and Astronomy, Northwestern University

"Electromagnetism: Nature's Force that Shapes our Lives is an inexhaustible source of learning and pleasure. Prof. Fagg undertakes to introduce us to the most mundane of Nature's four basic forces, namely, electromagnetism. This he does with erudition, nontechnical style, pedagogical skills and great care of details. The result is a popular book that every educated person can enjoy, as well as students, undergraduate and even graduate, and their professors.

All in all, Electromagnetism provides a wonderful voyage to the frontiers of physics and philosophy. The author brings to this voyage not only his vast knowledge acquired through many years of research and teaching in physics, but also the wisdom, depth, life experience and wit earned with age. I find this book a precious gift to any person who loves science."

Avshalom C Elitzur, Ph.D.

Iyar, The Israeli Institute for Advanced Research, Rehovot, Israel

FOREWORD

Aside from flipping the switch or resorting to a flashlight in a dark room, we in the modern world use a thousand gadgets that use electricity. But not many of us are aware that electricity is intertwined with magnetism, the glue that helps us display items on the refrigerator door. The undergirding principle of electromagnetism plays an indispensable role in virtually everything we see and experience as human beings in this physical world that is wrought with so much splendor and wonder. Not just the materials we find around us but our own bodies too are such as they are because of electromagnetic forces. More than that: every aspect of our perception from sight and smell to hearing, tasting, and touching, involve electricity. Even thoughts emerge from our brains whose neurons carry electric current also. Electromagnetism is an invisible undergirding principle that sustains the whole universe such as it is.

This book is a treasure chest in the enormous breadth in the topics it covers. Ubiquitous electromagnetism is the central topic, and we learn about its enormous impact on a variety of aspects of world. The author talks about the macroscopic world and also about the microcosmic. If the earth is a huge magnet, atoms and molecules carry electricity also. The book informs us about the great James Clerk Maxwell who gave the known laws of electromagnetism a unified mathematical formulation from which he made what may well be regarded as the most revolutionary scientific discovery in humanity's history in terms of its transforming power on modern civilization: the discovery of electromagnetic waves.

These pages also present in intelligible terms some of the most esoteric fields of modern physics, without burdening the reader with mathematical symbols. Einstein's famous formula connecting matter and energy is the only representative of the horde of mathematical jargon that clutter technical treatises on physics. Still we get a taste of the quantum theory of light and of quantum electrodynamics also. We read about the connection of electromagnetism to evolution, and we become aware of practical questions such as time measurement and GPS. Other matters explored here are electric power and computer chips.

It is remarkable that such a wealth of knowledge has been compressed within so few pages, and with such clarity in an engaging style. What adds to the interest of the book are the reflections of a very thoughtful physicist and his insightful comments sprinkled all through the text. The work reflects the author's extensive knowledge of and involvement with technical physics, his enormous respect for the grandeur of science, and also deep humility that befits a man of wisdom.

V. V. Raman
Emeritus Professor of Physics and Humanity
Rochester Institute of Technology
July 8, 2011

PREFACE

Especially in the last three decades there have been a considerable number of excellent books by distinguished physicists, astronomers, and cosmologists written to explain and popularize what has been learned about the physical universe. They have been primarily devoted to describing on the one hand the world of elementary particles: quarks, high energy particle accelerators, and theories of everything; or on the other hand, the physics of the cosmos: stars, galaxies, the big bang, and the expansion of the universe. Indeed some of these authors have been able to discuss both worlds with elegance and clarity.

In this book I focus attention on the physics underlying the world of our more direct experience right here at home on Earth. More specifically, my writing represents a sincere attempt to show how one of nature's four forces, the electromagnetic force, is the physical basis for almost all of what goes on in the vibrant life of this planet. I am driven by an unrestrained wonder at how this one force can do so much, a wonder that has fueled an ardent desire to share what I feel about what I have learned.

However, any attempt to write a single book about the universal role that electromagnetic phenomena play in the operation of earthly life is destined to be an incomplete undertaking. Nature exhausts our capacity to describe it. Nevertheless the challenge is there. So I have found that all I have been able to do is to draw on what knowledge I have gained through experience and study and present a glimpse at how this one force of nature dynamically undergirds our lives, our technology, and the nature that surrounds us.

Given this daunting mission, it must be emphasized that this is not a physics textbook on electromagnetism. It is a book addressed to the interested lay reader. While I do try to describe some of the principal characteristics of electromagnetism, I do so without mathematical equations with the exception of Einstein's famous $E=mc^2$.

The phenomena of nature are so incredibly diverse that the best I can do is to give an occasional glimpse of how the electromagnetic force is utilized in a few specific cases. So for the most part I feel that it is most appropriate to present a panorama, a landscape of the nature that so universally harnesses the electromagnetic force, with the primary intent of conveying the wonder of the role it plays in earthly life. In part because I am a physicist, the treatment of electromagnetism's function in technology, is a somewhat more manageable task. But even here the best I can do is to describe some representative examples.

So this book is my attempt to share my enveloping sense of awe about electromagnetism's underlying omnipresence in our world, an omnipresence that affords us another, deeper dimension in understanding how intimately we relate to this earth. I share this with the hope that others may be inspired to expand on, and refine, what I have presented here, which is only an introductory outline of a vast subject.

I begin the book with the story of how I learned about electromagnetism which will in turn give insight into how I view the subject. In the first chapter I discuss how electromagnetism came to be in the universe's evolution and how it has left abiding imprints on the cosmos in terms of the stars, galaxies, and the cosmic microwave background. This is followed by a portrayal of this world's nature that is everywhere vitalized by electromagnetism.

The next three chapters are introduced with a brief outline of the history of our understanding of electromagnetism, but are primarily devoted to a description of the basic characteristics and operation of the electromagnetic force. In the sixth chapter I discuss the universal role that this force has in modern technology through the use of selected illustrative examples. In Chapter 7 I describe how indispensable electromagnetism is in observing, and informing us of, the other three forces of nature: the strong nuclear, weak nuclear, and gravitational. In Chapter 8 I show how it informs us about our concept and measure of time, and I conclude with a chapter sharing my thoughts and contemplation about the significance of electromagnetism in our life.

I am deeply indebted to Dr. Joe Rosen for his thorough, meticulous reading and critique of every chapter of the book. I also wish to express my appreciation to Dr. J. R. Leibowitz for his helpful comments on several parts of the book, to Ms. Nancy Monacelli for her work on the figures, and to the staff at the Niels Bohr Library of the American Institute of Physics for their excellent reference work. The abiding support of my deceased wife, Mary Godfrey Skipp, along with her many helpful suggestions, I will remember with heartfelt gratitude for the rest of my life.

Introduction

Learning about Electromagnetism

I live on a small farm in the Shenandoah Valley of Virginia. Although quite modest in size, it is agreeably balanced with some patches of woods and some gently rolling pasture. Thus there is enough variety so that, before she died, my wife, Mary, and I were never bored taking short walks to different parts of the farm.

One time, shortly before sunset, while we were walking just below a gradual rise in the pasture, I happened to look over at the setting sun as its rays grazed the broad crest of the rise and was amazed at what I saw. The sun's rays had highlighted myriads of spider web strands that linked blades of pasture grass with a golden iridescence. The strands linking the grass blades pointed in every which direction so that the entire surface of the pasture was in effect covered with a soft gossamer blanket of spider web strands we never knew were there. It was a serendipitous glance at the vibrant mini-jungle that thrived right under our feet.

This experience set off a train of thought that in some indefinable way brought a sense of integration and completion to my view of the world around me. This was because, first, I realized that there was a still deeper realm of activity underlying this blanketing mini ecosystem. It was the micro world of the electromagnetically interacting atoms and molecules that are the fundamental ingredients of the spider web strands and the blades of grass. There was a ceaseless atomic activity "underneath" this blanket on which we were treading. Second, the sunlight that nourished the grass by photosynthesis and made it possible to observe this phenomenon was electromagnetic radiation from the sun's intensely hot surface.

Thus in one cohesive realization I was able to sense the micro-world of atoms and molecules, the macro-world of the spider web blanket, and the cosmic world of the sun, each involving the electromagnetic force in one way or another. What I had learned over the years in my study of physics, and specifically about electromagnetism, certainly informed my ability to come upon this perception.

The perception exemplified a growing engagement and fascination with the depth and scope of electromagnetic phenomena in our world that became especially apparent to me in the last fifteen years. On further thought, however, as I look back on my life, I realize that this view was the result of a lifelong incremental process of conscious or unconscious learning.

Indeed, I believe it started some 65 years ago after finishing undergraduate school. During the ensuing year I wandered through a series of jobs, including testing electrical equipment for General Electric Company and teaching mathematics at a prep school, but ended the year as an assistant engineer, testing model airplanes in a wind tunnel for the US Navy.

This experience sparked the idea that I wanted to get a master's degree in aeronautical engineering. Accordingly, as autumn approached, I consulted the Aeronautical Engineering Department at the University of Maryland. I was informed that before I could start a master's program, I would be required to take a number of additional undergraduate courses. I rebelled at this. I decided that I wanted a masters degree in something technical or scientific, it didn't matter what. So I decided to go to the Physics Department, where they made no such requirements, and I promptly decided to study physics.

In effect I stumbled into physics like a blind man walking down a road full of potholes. Reflecting on it now, although I would be hard pressed to think of a more trivial reason for starting such a career, I have never regretted it.

After a year of graduate school at the University of Maryland, I spent a year at the University of Illinois, where I was introduced to nuclear physics, the study of which I also have never regretted pursuing. I ended up getting my PhD degree in nuclear physics at Johns Hopkins University, where my thesis project involved the use of a small particle accelerator.

This early work set my direction in nuclear physics. For the rest of my career I was involved in experiments at a number of different laboratories in the US and Europe, where some property of the electromagnetic force was utilized to accelerate an electrically charged particle, such as a proton or an electron, in order to study characteristics of various nuclei. Some property of electromagnetism was again utilized in energizing large magnetic spectrometers and complex electronic equipment that were necessary for the detection and sorting of the particles emitted by the nuclei after exposure to the beam of accelerated particles.

But in order to understand what the experimental data were revealing about the nuclear properties, the analysis of the data necessarily required a full understanding of the electromagnetic interaction between the accelerated charged particle and the electric charges and currents in the nucleus. Thus, knowledge of electromagnetic interactions was necessary in the acceleration of the charged particles, in the detection of particles emitted by the nucleus, and in the analysis of what was detected.

Realizing how absolutely essential the electromagnetic force was in every stage of my nuclear physics work brought about a gradual transformation in my interest. That is, although the study of the nucleus continued to concern me, I became increasingly interested in the electromagnetic force itself. In effect I found myself as interested in the means of performing the experiment as I was in its end result. Concurrently, I began to see the extent to which electromagnetic phenomena came into play in the world outside of the laboratory.

I recall one particular event some twenty five years ago that nurtured my thinking about electromagnetism from this broader point of view. This came late in my career, while I was a research professor in the Physics Department at the Catholic University of America. I was asked to give a lecture on electromagnetism to an undergraduate class, substituting for one of the professors who had to be away.

I took the task seriously and spent considerable time in preparation. The class went well, well enough so that at the end of the lecture I was somehow inspired to say what had been stirring in the back of my mind for a long time. I pointed out how universal the electromagnetic force was in activating all of chemistry and biology. I went on to describe what that implied for the living earth as well as virtually all of modern technology and our capacity to observe the heavens.

As best I can remember, this was the first time that I verbalized publicly my thoughts about electromagnetism's omnipresence. I look upon that experience as seminal in the evolution of my near passionate desire to express how astonishingly extensive is the activity of the electromagnetic force on this living planet.

This general perception is certainly not new. In the seventeenth century, long before James Clerk Maxwell showed that electricity and magnetism were aspects of one force, electromagnetism, Theophilus Gale began his book with:

"Electricity the principal agent of animal life-of the
vegetable life.....this latent, mysterious and powerful
agent, pervades all creation, is capable of assuming
such a variety of appearances, and producing such a
variety of effects, both in the animate and inanimate
creation,..."

(Gale, published in 1802)

In about 1850 a somewhat similar viewpoint, but including magnetism, was held by Karl Freiherr von Reichenbach and William Gregory (von Reichenbach and Gregory, 1850; von Reichenbach, 1851).

I am building on this legacy from the past when I say that it is the pervasiveness of electromagnetic phenomena functioning in every organ in our bodies, in the life of the vibrant nature enfolding us, as well as in modern technology that continues to evoke in me a profound sense of wonder. I also wonder at how these phenomena operated at the very core of the continual, often exquisitely sensitive, process of trial and error during the eons of our evolution and that of living nature.

Thus the electromagnetic force was present to be brought into use early on and played an indispensable role in the evolution of the earth itself. This realization suggests looking back even further in time to see what current knowledge of cosmology can tell us of the origin of this force and its relation to the other three forces of nature. What markers did this force leave to allow us to know about its history? How did this force emerge from its grounding with the other three forces to make possible this living earth, this thus far unique flower in the cosmos?

CHAPTER ONE

THE EVOLUTION OF ELECTROMAGNETISM

On some moonless nights Mary and I used to see the stars with almost unimpaired clarity. Outside of the upstairs bedroom, there is a deck, completely open to the sky. Since the house is roughly ten miles from any town of significant size, the night sky is not obscured by light contamination. Nor is its enveloping silence often disturbed by any nearby noise. The stillness of the sky, randomly scattered with myriads of stars, is mute testimony to the awesome depth of the firmament.

It is the light from the stars and galaxies that tells us about the extent of this depth. The finite and measurable speed of light sets the pace at which we learn about the behavior of the far reaches of the cosmos. Some of the farthest galaxies are observed to be over thirteen billion light years away. (A light year is the distance traveled by light in one year.) This means that the light arriving at the astronomers' telescopes today allows them to see the galaxies as they were over thirteen billion years ago. So the farther away a galaxy is the further back in time is our observation. Thus the history of the entire universe is spread out before our eyes, and it is electromagnetic radiation that tells the story.

Part of this story is about the birth of electromagnetism itself, which is deeply imbedded in that of the universe's early evolution. Some of the basic features of this evolution are described by what is known as the Big Bang Theory. However, in 1980 this theory was refined and subsumed by the Inflationary Universe Theory proposed by Alan Guth. Its name derives from Guth's ingenious idea that the universe underwent an exponential expansion in the very initial instants of its life. The theory explains a number of characteristics of the universe that its predecessor could not (Guth, 1997).

In the last several decades the inflation theory itself has undergone a number of refinements and improvements, and is generally accepted by most cosmologists today. This theoretical work has given additional stimulation to an already robust study of the universe by astrophysicists and cosmologists.

1

Using knowledge gained from astronomical observations in conjunction with that found from very high energy particle accelerator experiments, cosmologists can now say that the universe originated 13.73 billion years ago (Cowen, 2010). At its birth the universe was so compacted that it was some 10^{-33} centimeters (a billion trillion trillionth of a centimeter) in size. At this initial stage of the universe's life all the forces of nature are considered to have been unified and indistinguishable.

As the universe expanded from this extremely compact state, these forces separated step by step and became distinguished into the four forces we know today: strong nuclear (often simply called nuclear), electromagnetic, weak nuclear (often simply called weak), and gravitational. With this expansion the universe was concurrently cooling into a series of successively lower temperature "soups," each new "soup" containing more complex particles synthesized from the more elementary particles of the previous soups (Pagels 1985; Hawking 1988; Lederman and Schramm 1989). For example, when the universe expanded and cooled enough, the early soup of quarks and electrons condensed to a soup of protons, neutrons, and electrons, which in turn "froze out" to a soup of electrons and light nuclei, like helium.

Thus, in contrast to the particles coming together, the forces separated. The final step in force separation occurred at about 10^{-10} seconds (one ten billionth of a second) after the big bang, This was when the electromagnetic force separated from the weak force setting the stage for the formation of the last soup which is of special interest to us here.

ELECTROMAGNETISM'S FIRST IMPRINT ON THE UNIVERSE

It was with the formation of this soup at about 380 thousand years after the bang that electromagnetism left its first indelible mark on the universe. By this time the universe had expanded and cooled sufficiently to allow negatively charged electrons to attach themselves to positively charged nuclei by means of the electromagnetic force to form electrically neutral atoms, mostly hydrogen atoms. This made it possible for light, which is electromagnetic radiation, to move freely throughout the universe without being ensnared in constant interaction with the ionized medium of electrically charged particles that characterized the previous higher temperature soups. (Along with many physicists, I consider light to be all electromagnetic radiation, not just visible light).

Over the following billions of years as the universe expanded and cooled, the initially higher frequency and shorter wavelength of this radiation progressively changed to the very low frequency and very long wavelength it has today. The frequency and wavelength of electromagnetic radiation are directly related, because the product of the two yields the speed of light. It travels a distance of one wavelength in one cycle of the frequency. It is the cooled remnant radiation present today that constitutes what is called the cosmic microwave background (CMB). This background radiation surrounds us everywhere; it silently pervades the entire universe as an abiding electromagnetic legacy.

The discovery of the CMB in 1965 by Arno Penzias and Robert Wilson (for which they received the Nobel Prize in physics) provided a crucial part of the experimental support for the original Big bang Theory. The CMB was one of the features predicted in the first mathematical formulation of the theory by George Gamow and his collaborators.

The theory predicted that the CMB would have a particular variation of intensity of the radiation as the wavelength varied over a full spectrum, or range, of wavelengths. It was Max Planck who first discovered the mathematical expression for this kind of spectrum. To do this he had to introduce the concept of light as a particle, not a wave. This was a seminal step in the development of the quantum theory (See Chapter 4).

To understand the characteristics of this spectrum, we must remember that the CMB radiation was emitted when the universe was virtually a homogeneous hot body confined by the universe's space as it existed at age 380,000. The conditions are similar to that of the radiation emerging from a coal furnace. The spectrum of the emitted radiation can be presented as a plot of radiation intensity versus the wavelength. The plot shows a continuous curve that looks somewhat like a lopsided bell curve, leaning toward the shorter wavelengths as in Figure 1.1. Where the curve peaks, depends on the temperature of the emitting body, peaking at shorter wave lengths for higher temperatures. So although all material bodies at temperatures above absolute zero emit electromagnetic radiation, even our own, the CMB radiation has the particular characteristics shown in the figure.

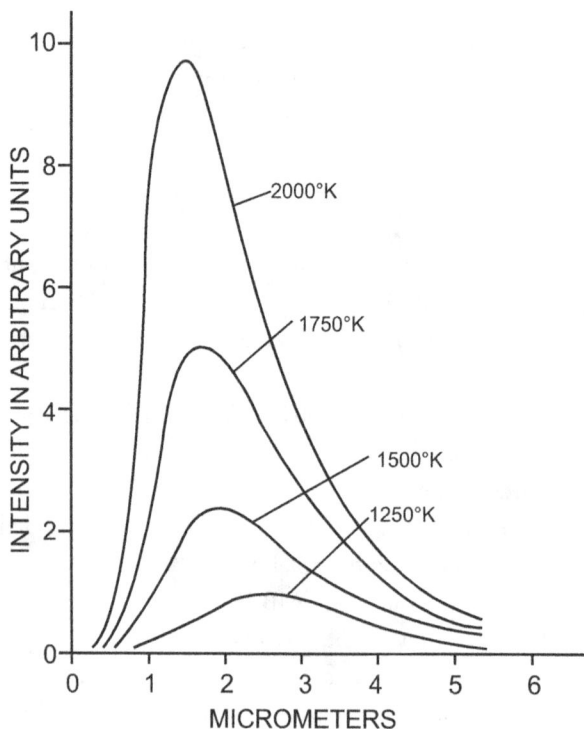

Figure 1.1 Radiation spectra from homogenous hot body at several different temperatures. The wavelength is denoted by λ. The temperatures are given in Kelvin degrees, the number of degrees above absolute zero.

I remember when I was attending a session of a meeting of the American Physical Society and the spectrum of the CMB observed by the COBE Satellite was shown for the first time on the screen. A groundswell of oohs and aahs swept the auditorium because the spectrum matched so beautifully that first derived by Planck. Indeed the curve matched to within one part in ten thousand thus constituting one of the major experimental supports for the Big Bang Theory.

Satellite and balloon observations soon to follow, however, showed that the CMB contained a wealth of further information about the behavior of the early universe. For example, finer resolution observations of the CMB by the Wilkinson Microwave Anisotropy Survey (WMAP) at the level of almost one part in one million revealed more accurately temperature variations in the radiation. The variations appear as a mottled array of blotches of higher and lower temperature regions as shown in Figure 1.2. The blotches are a primordial electromagnetic blueprint for the structure and distribution of galactic clusters we see today (Schwarzschild, 2006). The lower temperature regions correspond to more dense parts of the universe which, through gravitational attraction over the eons, pulled in more and more matter to become the galactic clusters observed in the sky today.

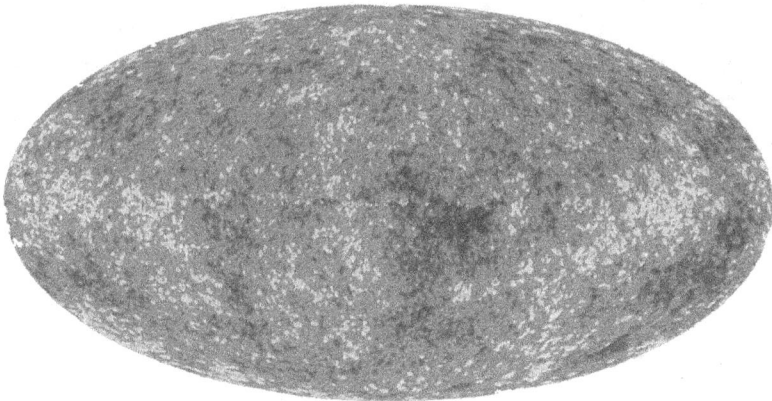

Figure 1.2 Cosmic microwave background radiation as observed by the WMAP satellite. The darker patches correspond to regions of lower temperature.

A meticulous analysis of the size and spacing of the temperature variations revealed by the WMAP has yielded results that support the Inflationary Universe Theory. For example, one such result is that the universe has a flat geometry. To understand this, let us imagine that three stations in space have been established millions of miles apart. The stations can send and receive light signals to and from each other. They are situated so that they form a triangle. If we add up the angles of the triangle and they sum to 180 degrees, as they do from our experience in plane geometry, the universe's space is flat, which was found to be the case. Otherwise, the sides are either curved out, as if drawing the triangle on a globe, or curved in, as if drawing it on a saddle.

Other meticulous analysis of the CMB's electromagnetic blueprint has provided information about the percentages of matter and energy in the universe today. The study of the CMB is an on going effort which will continue for years. For example, the Planck satellite probe launched in 2009 is expected to yield an even finer definition picture of the CMB. It is clear that what cosmologists and astrophysicists are doing is a prototypical example of milking the maximum amount of data from a phenomenon under study. The phenomenon here is the cosmic microwave background. It is the first and still abiding fingerprint of electromagnetic radiation that quietly permeates the cosmos and tells the story of the early universe.

A central part of this story is, again, that the gossamer-like mosaic of temperature variations are sources of minute density concentrations, seed attractors, which by means the gravitational force accreted more and more matter to become the clusters of galaxies we see in the sky today. The galactic clusters and the galaxies and the stars they contain grace the heavens with billions of points of light and constitute electromagnetism's second lasting imprint on the universe.

ELECTROMAGNETISM'S SECOND IMPRINT ON THE UNIVERSE

Virtually everything we learn about the stars and galaxies is made possible by means of light, electromagnetic radiation. As shown in Figure 1.3, it comes in all wavelengths from the shortest wavelength gamma rays through ultraviolet, visible, and infrared, to microwave and the longest wavelength radio waves. The center of the visible spectrum is in the yellow region roughly at $6x10^{-7}$ meters, and inspection of the figure shows that this corresponds to a frequency of about $5x10^{14}$ cycles per second. The product of the two numbers is $3x10^8$ meters per second, which is the speed of light.

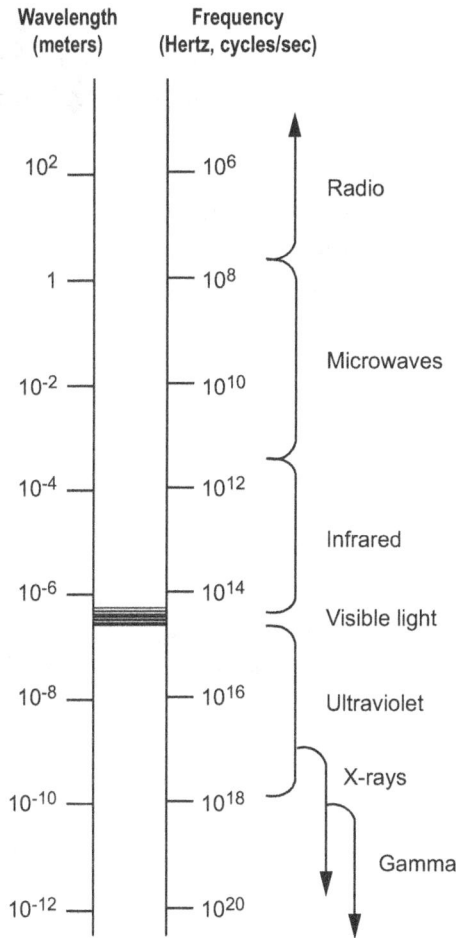

Figure 1.3 The spectrum of electromagnetic radiation given in terms of frequency and wavelength.

The light from the stars is the result of the tremendous heat generated by nuclear reactions, whose outward pressure resists the inward pressure of gravity. Stars come in a range of masses and spectra of light they emit. The brightness, or luminosity, of a star depends directly on its temperature. A plot of the luminosity of observed stars versus their temperature shows that most stars generally cluster along a line extending from the hottest, brightest stars to the coolest and dimmest and form what is called the main sequence. Our sun is a little beyond midway in this sequence. The plot is known as the Hertzsprung-Russell diagram shown in Figure 1.4. The hottest stars tend to preferentially emit blue light and the coolest, red light, just the opposite of what you see on some bathroom spigot knobs. A small percentage of the stars, however, known as red giants, near the end of their life, and others called white dwarfs, at the end of their life, are outside of the main sequence.

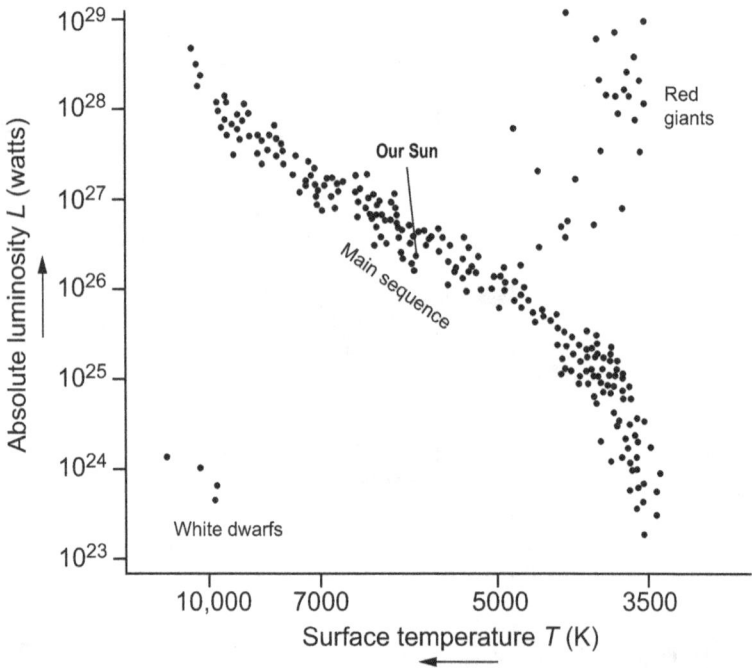

Figure 1.4 The Hertzsprung-Russell diagram of the luminosity of stars plotted versus their temperature. Most stars are located along the main sequence from upper left to lower right. The sun is indicated near the centre of the sequence.

By examining the spectra of light a star emits, astronomers have determined that all stars consist largely of hydrogen and helium. However, all stars are ultimately doomed to collapse due to the relentless pressure of gravity and the finiteness of any source of energy (Pagels, 1985, p.30). It is in the nuclear burning in the stars that the heavier elements are made and then thrown into interstellar space when the star collapses. Second generation stars then form that contain these elements, the "star-dust" of which we ourselves are made.

Astrophysicists have determined that there are three possible destinies for stars collapsing at the end of their life. Depending primarily on their mass, they can become white dwarfs, neutron stars, or black holes. Those whose mass is less than 1.4 times that of the sun collapse to white dwarfs, small but with mass densities thousands of times greater than ordinary matter. Stars more massive than this limit with masses as great as about three suns, suffer a supernova explosion leaving behind a neutron star.

Neutron stars are so compressed that the electron that would ordinarily swirl around the proton nucleus in a hydrogen atom, for example, is forced to join the proton nucleus to become a neutron. The result for the star is an incredibly dense compaction of neutrons the size of a city. In fact neutron stars are so heavy that one teaspoonful of the star weighs roughly a billion tons. A more catastrophic fate is in store for even more massive stars that implode and collapse into black holes, so named because not even light can escape their overwhelming gravitational pull.

However, among the vast diversity of stellar phenomena about which electromagnetic radiation informs us, perhaps the most impressive is a certain type of neutron star known as a pulsar. In 1967 Jocelyn Bell, a graduate student at Cambridge University, noticed a weak but repetitive signal on the rolls of paper recording data from observations of certain parts of the sky (Layser, 1984, p.219). Anthony Hewish, Bell's thesis adviser, checked the data and found that the timing of the signal was accurate to one part in ten million. A number of causes were considered, including extraterrestrial intelligence. After several other similar signals were discovered in different parts of the sky, it was concluded that they came from the rotation of neutron stars with very strong magnetic fields (Pagels 1985, p.64).

When the north-south polar axis of the magnetic field is not pointed in the same direction as the axis of rotation of the star, the result is pulsing electromagnetic radiation. In effect it is a "stellar lighthouse," whose timing precision is truly remarkable for such a massive object. The high rotation rate is due to an effect that is entirely similar to a spinning ice skater when she pulls her arms in. So when the star collapses to such a small volume, its spin increases enormously. Pulsars are huge, churning electromagnetic engines that call into play the full array of electric and magnetic properties

Neutron stars have received special attention in recent years for another reason. The light resulting from the supernova collapse to such a star has played a vital role in our understanding the dynamics of the cosmos. A particular type of this phenomenon is known as a supernova 1a. When two stars rotate around each other, one of which is a white dwarf, the much denser white dwarf drains matter from its companion. This process continues until its mass reaches the critical limit of 1.4 solar masses, at which time it explodes in a supernova leaving behind a neutron star. Since the mass accretion rate is so steady and reliable, the explosion occurs quite accurately at this limit.

This means that all supernova 1a explosions produce essentially the same amount of light. Having such uniform luminosity, they are known as "standard candles" by astrophysicists. As such they can by used to determine distances. Since the measurable, or apparent, luminosity of a light diminishes with the square of the distance from the observer (that is, decreasing a factor of four at twice the distance), a simple calculation yields the distance from the supernova.

In 1998 astronomers noticed that supernova 1a explosions in some distant galaxies were fainter than expected, given what was then known about the universe's expansion rate. It was soon realized that the universe was expanding at an accelerating rate. Because of this, it is predicted that eons from now, for example, we will be able to see only about six nearby galaxies (Krauss, private communication).

It is somehow comforting to know that at least for a very long time galaxies will be observable. Galaxies are enormous star making machines, nurseries for all varieties of stars born out of interstellar gas. The distribution and configuration of the gas clouds as revealed by the Hubble Space Telescope is truly awesome. The variety of colors from every portion of the visible electromagnetic spectrum should be a challenge and inspiration to any abstract artist. Even more engaging is the fact that images of a given cloud system can be seen to have quite different configurations, depending on whether they are observed by an infrared, visible, or ultraviolet telescope.

There are some one hundred billion galaxies in the observable universe, each containing on average some one hundred billion stars. As far as we are able to perceive now, in all of this vastness there is only one galaxy containing one star, the Sun, which is very special. It is so special because it harbors and nurtures a system of planets that includes the Earth, which so far as we know now is one of a kind.

ELECTROMAGNETISM'S THIRD IMPRINT ON THE UNIVERSE

This is because there are extraordinary properties of the Earth and its place in the solar system that are uniquely favorable to life. Its orbit is essentially circular with a distance from the sun that allows an appropriate life-supporting temperature range. The angle of its axis of rotation with respect to axis of the orbital plane is stabilized by a moon of appropriate mass, thus giving us the regularity of the seasons. Jupiter's large mass keeps most destructive comets and asteroids out of Earth's path. Experts in planetary motion tell us that such a remarkable combination of these characteristics is a very rare phenomenon.

The list of exceptional circumstances goes on: from the existence and movement of tectonic plates, giving us our supply of iron and other metals for our technology, to the timing of the appearance of water and oxygen, to the global extinctions, including that of the dinosaurs, making possible the incredible biological evolution of plants, animals and humans. After the discovery of almost two thousand extra-solar planets, mostly by the Kepler satellite (as of February 2011), none come close to the distinctive features of the Earth.

Moreover, using selected parts of the electromagnetic spectrum, astronomers have been looking for signals indicating the existence of extraterrestrial intelligence for 50 years with no success. While it is very important that they continue to look, their effort is encumbered by the finiteness speed of light. Though such intelligence may exist somewhere in the universe, the isolation imposed not only by light's finite speed, but also by the finiteness of our lives dictates that we may never really know.

These are but a few of the reasons why the Earth seems to be singular. But a consummate reason for its singularity is that it has been where electromagnetism has been free to pursue its rich potential, a potential realized in the evolution of human intelligence. The Earth, this flower in the cosmos, hosts the myriad of electromagnetic phenomena that make possible virtually all of our life. To understand why this is so we must explore the incredible multiplicity of roles that these phenomena play on this planet in order to fully appreciate their wonder.

CHAPTER TWO

OUR ELECTROMAGNETIC EARTH

Many times after I have finished breakfast, I spend a few pensive moments gazing out the glass doors half consciously curious about what the birds and squirrels are up to. Also, since we have not allowed hunting on the farm, the deer can occasionally be seen enjoying some salad from the flower garden. But sometimes when there is only a hint of a breeze, I may happen to look at the silver maple tree behind the bird feeder and my attention will be taken by one of its leaves. A local confluence of the breeze in the tree, a kind of quiet mini vortex of air, will cause that particular leaf to wave back and forth. In an admittedly childlike, but enjoyable, moment I will feel that the leaf is waving at me, saying good morning.

When I reflect on such moments, I sense how the leaf enjoys its vitality by means of the veins that serve as the skeleton for its symmetric shape, veins that deliver the nutrients coursing up from the earth and through the tree trunk and branches. Trees grow by means of the cellular activity concentrated in what are called apical meristems, located at the tips of the branches and roots. It is electromagnetic interactions that drive this cellular growth.

Drawing on the physicist in me, I realize that this is an example of how all of the nature that I see utilizes the electromagnetic force for its activity. It is here on this lonely planet where the universe's consummate evolution has proceeded by grace of electromagnetism's rich, dynamic potential. Its effect and presence in all aspects of our life and relation to the world is ubiquitous.

The number of living creatures and other natural earthly phenomena where it can be shown that this is true is virtually infinite. So, given the scope of this book, it is not possible to describe in detail, starting with the sub-cellular level of a tree, for example, how electromagnetism functions in its roots, its trunk, and its branches and leaves. Such an undertaking for all of the earth's natural fecundity would involve endless volumes of meticulous bioelectric discussion.

Thus, although it is admittedly only scratching the surface, I have chosen to give some idea of the scope of these phenomena by first discussing how the electromagnetic force operates at the atomic and molecular level underlying all of the earth's nature. Then I wish to convey the breadth of its plenitude by presenting a panorama in terms of a number of representative examples, and in selected cases by giving a glimpse of electromagnetism's role in the example.

THE MICROSCOPIC WORLD

Let us start by looking inward at the microscopic world of atoms. Negatively charged electrons are constrained to swirl around the positively charged nucleus of an atom in a kind of quantum cloud by the electromagnetic force. The quantum refinement of electromagnetic theory shows that the force is considered to be transmitted by undetectable particles called virtual photons in contrast to the real photons of light that make it possible for us to see the world around us (see Chapter 4).

It is this same interactive photonic "glue" that holds atoms together in a molecule so that all of chemistry and biology operate fundamentally by means of electromagnetic interactions. Let us see how this is true for the two molecules that are prime contents of the earth's oceans.

Common table salt, or sodium chloride, affords an especially clear example of simple molecular binding. This binding can be easily understood by examining what the quantum theory tells us about atomic structure. According to the theory, the electrons in the atom, which electrically balance the protons in the nucleus of the atom, are arranged in concentric "shells" of electrons. Each successive shell contains more electrons than the previous one. In the case of the sodium atom, it has a completed shell of electrons plus one in the next shell. On the other hand the chlorine atom has a completed shell minus one electron. Consequently, there is a significant tendency for the sodium atom with its outer electron to be electrically attracted to the chlorine and fill the vacancy in its outer shell and form a salt molecule.

Because of its indispensable necessity for life in the oceans as well as that on land, the water molecule is especially important to consider. The properties of this molecule of two hydrogen and one oxygen atom are so numerous and varied that it is still under study to this day. The outer shell of the oxygen atom has two vacancies which are filled by the electrons of the two hydrogen atoms to form a triangular configuration. This configuration allows the assembly of triangular molecules as building blocks for the formation of the incredible variety of beautiful hexagonally symmetric snow flakes. The crystallization process that occurs when water freezes to ice at zero degrees centigrade causes the ice to become less dense than liquid water, thus making it possible for fish to live underneath the frozen surface of a pond in the winter.

The salt and water molecules are but simple examples of the vast number of often very complex molecules that make up the study of chemistry. But in all cases the basic force employed in binding the atoms together to form the molecules is electromagnetic.

ELECTROMAGNETISM, THE WORKHORSE OF EVOLUTION

Indeed considering the still higher levels of complexity evident in biological processes, we realize that our very evolution has depended on a multitude of electromagnetic quantum interactions, largely of exquisite sensitivity. They are put into service for the incessant probing, trial-and-error process of communicating, attracting, and repelling that has made possible each new level of complexity.

Evidence for how the first level may have arisen was revealed in the seminal experiment of Stanley Miller, then a graduate student of the distinguished chemist, Harold Urey. He filled a flask with a mixture of organic gases such as methane and carbon dioxide along with ammonia to approximate what was thought to be the earth's atmosphere 3.5 to 4 billion years ago. The gaseous mixture was then subjected to a series of sparks to replicate lightning. Several days later, he observed a number of different amino acids, building blocks of proteins, clinging to the bottom and sides of the flask.

The chemical reactions of the gases all proceeded by means of electromagnetic interaction and the lightning is one of nature's most explicit displays of its latent electric power.

This experiment served as a relatively simple illustration of how there is a thrust toward molecular complexity in all of nature. The most impressive aspect of this thrust is apparent when we realize that it was from such primitive beginnings that proteins and enzymes, molecules of far greater complexity, evolved (Ricardo and Szostak 2009, p. 54). Stuart Kaufman in his extensive study on this subject speaks of "the profound power of self-organization in complex systems" (Kaufman 1995, p.43). But these systems are the product of what he calls molecular creativity, the capacity to create complex molecules (Kaufman 1995, p. 132). This creativity demonstrates how bioelectricity is a feature of all living things.

So it was that the unrelenting, painfully slow, process of experimentation and adaptation over billions of years reached a milestone with the formation of cells, the primal components of all living creatures. Inside of the cell's protective membrane are chromosomes made up of genes. The genetic information in a gene is found in a DNA (deoxyribonucleic acid) molecule consisting of a long chain of smaller molecules that are called nucleotide bases, which come in four types: adenine (A), cytosine (C), guanine (G), and thymine (T). Two DNA molecules are entwined and electrically sustained as a double stranded helix as shown in Figure 2.1. Throughout each strand only an A base can be attracted to a T, or a C attracted to a G (Giancoli 2005, 460-2).

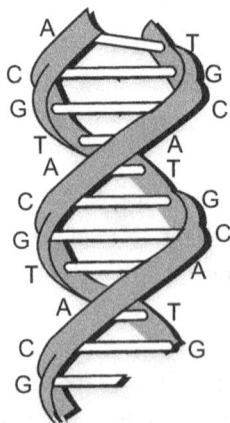

Figure 2.1 Diagram of a DNA double helix molecule. Note that only A and T bases and only C and G bases are linked to each other by means of electrostatic force.

©GIANCOLI, DOUGLAS C., PHYSICS: PRINCIPLES WITH APPLICATIONS VOLUME II (CH. 16-33), 6th,©2005. Printed and Electronically reproduced by permission of Pearson Education, Inc., Upper Saddle River, New Jersey.

16

This selective attraction is due to the unique shapes of the molecules, so that only the AT and CG pairs fit and can come together by electrostatic attraction, the same force that causes your hair to be attracted to your comb when both are dry. The electrostatic force between the pairs is weak and binding occurs only when there is a proper fit. This is why so few mistakes are made in the DNA's construction, which replicates with the help of enzyme molecules that also act via an electrostatic force. So amidst the random thermal motion of such organic molecules, it is the electrostatic force that "acts to bring order out of chaos" (Giancoli 2005, p. 462).

The DNA is the nucleus of the cell and contains its operating instructions. Other cell components are organelles, each with a specific task. With the "push-pull" of miniscule electrochemical gradients (rates of increase or decrease in electrostatic voltage with distance), organelles execute the instructions necessary for the cell's maintenance.

Equally tiny such electric signals come into play in the awesome collective behavior of bacteria, the smallest living cells. Coherent assemblies of bacteria exhibit a purposeful mobility and remarkable sophistication in their growth and survival. They are indeed survivors. To quote James A. Shapiro:"... bacteria are outstanding genetic engineers, and they have used this capacity to withstand antibiotic chemotherapy" (Shapiro, 1995).

The behavior of bacteria is an example of the viability of cellular life and interaction at a primitive level. But similar interactions prevail at every level of living complexity. The extreme subtlety of cellular interactions is quantified in experiments in microbiology which show that voltage gradients as low as one ten millionth of a volt per centimeter and frequencies between 0 and 100 cycles per second are involved in all living creatures. All plant and animal life is bathed in, and interacts with, a sea of such very low frequency radiation that envelopes the earth (Adey 93). This is essentially because any material body, human or otherwise, whose temperature is above absolute zero, radiates some low level electromagnetic radiation.

Cellular cooperation in plant life is silently proceeding around us all of the time. For example, it is electrochemical processes that do the work in the photosynthesis of plants and algae by harvesting sunlight's electromagnetic energy in order to conduct their metabolic reactions. These reactions produce the chlorophyll that clothes plants and trees with variegated shades of green (Hu and Sculten 1997).

Moreover, plants can employ extraordinary ingenuity in protecting themselves from predators or avoid competitors. By analyzing the ratio of shorter to longer wavelength red light shining on its leaves a plant can sense the presence of other competitors nearby and try to grow in another direction (Angier, 09). The smallest bite of a hungry insect causes certain cells on a plant's surface to release chemicals that irritate the predator, or a sticky substance to entrap it (Angier, 09). Some responses are remarkably fast. Experiments by Consuelo De Morales and her colleagues at Pennsylvania State University show that in less than 20 minutes from the time a caterpillar begins to feed on a plant, the plant, drawing on carbon from the air, begins to generate defensive compounds. Some compounds that plants generate are volatile chemicals that serve as alarm calls attracting such predators as dragon flies that are especially partial to caterpillar meat (Angier, 09).

DECEPTION IN EVOLUTION

The caterpillar is, of course, unaware of the release of the chemicals, so there is some deception at play. Indeed deception is a vital and fundamental feature of the entire process of evolution (Rue, 1994, p.108). Deception involving such tricks as camouflage and mimicry is practiced even at the level of cellular interaction in the human body. Viruses are able to generate a coating that either protects them from the attacking blood cells of the immune system, or simulates the coating to avoid detection.

Of the thousands of species of orchids, some are designed to emit molecules that mimic pheromones which are molecules that convey sexual signals. Thus insects interested in sex are tricked into pollinating the orchid. Indeed there are orchids especially designed to attract specific insects, e.g. a fly, bee, or spider (Rue, 1994, p. 110). But a crab spider can hide in between the leaves of a flower, or make its legs approximate the color of the flower and wait for its meal. In another tactic of deception, the larvae of the swallowtail butterfly appear very much like bird droppings which the birds naturally avoid. Moreover, when the larvae molt becoming a caterpillar, it is green, safely camouflaged on the leaf it is eating (Rue, 1994, p.112).

Television productions about life in the ocean have been especially effective in showing a striking arsenal of strategies of camouflage and mimicry that fish and other sea life practice in their struggle for survival. But an exceptional example, perhaps not often seen, where electromagnetic radiation is being put to consummate use, is the capacity of some deep sea fish to use light-producing organs called photophores, which are distributed along the belly of the fish. The amount and spectrum of light received by the upper half, or back, of the fish is processed, exactly reproduced, and emitted from its belly by the photophores. In effect this renders the fish virtually transparent. Indeed, the hypolyte prawn actually is transparent (Rue, 1994, p.116).

The use of deception also prevails throughout other members of the animal kingdom. Birds will flair out their wings to appear larger to a predator and deposit their eggs in the nests of other species in order to freeload the incubation of their offspring. A lower ranking chimpanzee will use one hand to conceal his erect penis from the head male, but leave it exposed to the nearby female with whom he wishes to copulate (Rue, 1994, p.123).

ELECTROMAGNETISM FOR BETTER OR FOR WORSE

The practice of deception by living creatures throughout evolutionary history informs us about a basic reality concerning the breadth of phenomena where the electromagnetic force finds utility. Indeed, while it is a force that all biologic species rely on for their growth and sustenance, it also is one that figures in their decay and death. The versatile properties of electromagnetic fields play their natural role in the decay of a tree as well as in the vicissitudes of our own aging. The death of the discarded vegetables in the compost bin, however, generates the rich soil that nourishes the new vegetables in the garden.

Thus electromagnetism is completely neutral on the question of how it is put to use. Adolph Hitler and St. Francis of Assisi both drew upon the similar bioelectric undercurrents. This is true also for the polio vaccine and the virus it fights. My wife and I have both had to deal with cancer. But when I can muster a dispassionate viewpoint, I realize that a cancer cell in itself is not evil; it is pursuing its course as a phenomenon of earthly nature. The same is true for the HIV virus, a malarial mosquito, or a deer tick.

The beautiful and beneficent, as well as the dangerous and deadly, live side by side. Sometimes they coalesce. The strikingly attractive colors of a coral snake are not an invitation to come too close. The sleek undulations of an electric eel come with a potential to deliver currents of one ampere at 600 to 1000 volts. The collective electromagnetic interactions of air and water molecules assemble to form a tornado that exhibits its own awesome symmetric beauty. The brilliant orange along the top of the poison arrow frog along with the engaging design of black and white over the rest of its body are a warning to stay away.

But it must be remembered that seeing these diverse natural phenomena as beautiful or dangerous (or beautiful and dangerous) is essentially a human viewpoint. The wonder is that humans over the eons have evolved in a way that they could have a viewpoint. It is humans, the most exquisitely complex of nature's fruits, that have the capacity to be aware of the nature from which they arose and by which they are nourished. It is humans that draw on a range of electromagnetic properties very intimately every moment of their daily living. This is so from the interactions of blood cells in the body to the activity of neurons in the brain that makes possible human consciousness.

WE AND THE WORLD

Among the most exquisitely sensitive human organs that help inform this consciousness are the eye and the ear. It is the eye that responds to nature's rich spectrum of colors and plethora of subtle hues that electromagnetic radiation transmits. The muscles controlling the shape of the eye's lens and the iris regulating the size of the pupil are enabled via electromagnetic interactions. The rods and cones of the retina triggered by the light focused by the lens send information through the optic nerve to the brain by means of electric impulses. However, these functions of the eye tell us that we cannot actually see light while it is in transit. As with any detector of light, only when the light interacts with the sensing components of the eye do we experience seeing the light.

Complementing the vista encompassed by the eye, and of comparable sensitivity, is the ear with its uncanny capacity to sense the immediacy and muted vibrancy of enfolding nature. The ear drum resonating to sound waves - waves of electromagnetically interacting molecules – transmits its message through the minute bones of the middle ear to the fluid of the cochlea, which in turn activates the sensory hair cells. These cells dispatch electric signals to the brain.

It is the human brain, however, that is evolution's consummate accomplishment. Nowhere in the known cosmos is the scope and potential of the electromagnetic force more completely utilized than in the incredibly complex "mini-universe" that is the human brain. It operates by means of some one hundred billion neurons, which is roughly the same number as the galaxies in the observable universe. Via electric signals fired across the synapses that separate them, a given neuron can receive information from as many as a thousand other neurons. From the most primitive functions of the cerebellum to the sophisticated operations of the cerebral cortex with its left and right hemispheres in continual communication, all depend on minute electromagnetic interactions. In fact, the adult brain generates its own electromagnetic activity known as the alpha-rhythm, which ranges roughly from 8 to 12 cycles per second (Whitrow 80, p.127).

Thus, the ability of your eyes to see and read this page and for your brain to process and interpret its contents depends on a host of what Geoffrey Chew calls "gentle quantum events" that vivify our bodies and give presence to our consciousness (Chew, 1985). These electrically driven events make possible the pleasure of seeing the whimsical flight of a butterfly, or two flirting with each other. The capacity of resonating strings to yield the soaring strains of Brahms's Violin Concerto, for the sound to travel to your ear, for your ear to transmit the sensations, and for your brain to respond to the music-all depend on such events.

The inanimate nature that we see: mountains, sea, and sky, and the wind and surf that we hear also depend for their characteristics on the variegated features of the electromagnetic force. One day after a rather heavy rainfall,

I stepped outside of the farmhouse and was stunned at the sight. It was the most complete and lucid rainbow I had ever seen. Because there are no buildings in sight, it scribed a full semicircle and each color, with the intermediate hues merging into the next, was brilliant. Myriads of raindrops serving as both refractors and reflectors had sorted out the colors of the sun's light. With this chance caprice of the weather, nature had vividly displayed a small portion of the electromagnetic spectrum for Mary and me to see.

Less dramatic, but quietly engaging, is the subdued dark blue of the venerable mountains in the distance that enclose the Shenandoah Valley. These are geologically old mountains that have settled into their gentle undulations for eons since they were first thrust upward by tectonic action. Even on parts of our farm there are outcroppings of rock on the slopes down to the creek that reveal almost vertical laminations that are hints of such ancient eruptions.

The apparently inert solidity of the rocks hides a vibrant stasis of molecules, molecules that are either geometrically arrayed in crystals or exist in an amorphous state. They are tightly bound by the virtual photon force carriers of electromagnetism, but held back from totally coalescing by the quantum dictum that only so many electrons can occupy a given atomic shell, all others are excluded. It is this same dynamic balance that keeps the table lamp from falling through the table and the table from falling through the floor. This restraining phenomenon that plays a crucial role in such a balance is called the Pauli Exclusion Principle. It was first enunciated by Wolfgang Pauli, one of the contributors to the quantum revolution early in the last century.

The geologic rock outcroppings also appear at certain locations along the creek and actually serve as small dams. It has been a true pleasure on occasion to rest a while beside the creek and listen to the water splash through a fissure in the dam. But the creek often disappoints and stops flowing in mid to late summer when dry weather generally prevails. This is when Mary and I used to visit her cousins, who were her childhood playmates, and live in eastern Maryland on the Sassafras River that quietly winds its way to the Chesapeake. Or we may have gone to the Delaware seashore.

The water molecules that collaborate to give us the gurgle of the creek, the wavelets that lick the shore of the river, and the surf that crashes onto the beach, are not restrained by a quantum exclusion phenomenon. They can yield the quiet trickle heard in a Japanese rock garden as well as the awesome force of a Pacific tsunami. But it is at the ocean shore where I can find an isolated stretch of the beach and experience a sense of openness to this earth, an openness that I can capture nowhere else. This is simply because it is open.

The blue sky that meets the ocean at the horizon imperceptibly changes the intensity of its azure hue during the day. It gets its blue because the molecules in the atmosphere preferentially scatter the blue (shorter wavelength) portion of the sun's electromagnetic spectrum. At sunset the sun's rays arrive at a glancing angle and pass through much more air, causing most of the blue to be scattered away, hence the redness.

The encompassing openness of the seashore setting can prompt my thoughts to contemplate an openness far more vast. We, and all living creatures, are fundamentally carbon based species. So let us consider the carbon atom. 99.97% of its mass is concentrated in the nucleus at its center and occupies some one trillionth of its volume. The rest of the volume consists of six electrons of very small mass and trillions of force-carrying virtual photons (discussed in Chapter 4) that keep the electrons in their orbits. Thus we are immersed in an ocean of unceasing photonic undercurrents that fill the vast majority of the world's space. In fact we are part of the ocean.

CHAPTER THREE

DISCOVERING ELECTROMAGNETISM

How do we know these things? How can I say that we are part of an electrodynamic ocean? I believe that it is primarily because we know more about the electromagnetic force than any of the other three forces. At a summer conference years ago David Park and I decided to have a picnic lunch together between conference sessions. A Professor Emeritus at Williams College, David is an excellent physicist and the author of many thoughtful and informative books about the natural world. I was musing about how we know more about electromagnetism than the other three forces, and David ventured that it might be because it is the force that we draw on for the operation of our bodies and brains. This somehow made sense to me when I thought of how immediate and intimate is our dependence on this force for our life.

Whether or not this interesting speculation has any validity, it is true that there are still unsolved problems with understanding the other three forces. For example, our understanding of gravity has been seriously challenged since the 1998 discovery that the universe is expanding at an accelerating rate due to an unknown repulsive force called dark energy.

In any case, it is certainly true that our extensive knowledge of electromagnetism is due to the genius and infectious curiosity of workers over a period of more than four hundred years. During about the first half of this period electricity and magnetism had been known as separate forces. Early in the nineteenth century, however, the accumulated work of experimenters and theoreticians showed that they were unified as two aspects of a single force. Here is a brief story of how it happened.

EARLY ELECTROMAGNETISM

Hans Christian Oersted, a professor of natural philosophy at the University of Copenhagen, had been pondering the relation between electricity and magnetism since about 1813. He had a persistent notion that somehow the two were related. In pursuit of this notion, one day in 1820 he demonstrated an experiment before his class that involved positioning a wire carrying an electric current in certain orientations with respect to a compass. The compass needle showed no convincing response. After class, however, he tried positioning the wire parallel to the needle and it swung through a large angle. When the current was reversed, it swung other way. (Purcell 1963, p. 148).

This observation triggered a series of experiments, the results of which he reported in an extended article titled "Electromagnetism" in the *Annals of Philosophy* in November 1821. Thus, for the first time a substantiated observation of a link between electricity and magnetism was made. This began the uncovering of a realm of physical nature the understanding of which has burgeoned to give us the principles that underlie the vast majority of phenomena on which modern technology is based.

On hearing of Oersted's discovery, André Marie Ampère immediately began his own investigations of the phenomena with a series of remarkably ingenious experiments between 1820 and 1825. These experiments led to a theory describing the characteristics of the magnetic strength in the vicinity of a wire carrying an electric current. James Clerk Maxwell (to be discussed shortly) later said that Ampere's work was "one of the most brilliant achievements in science." (Harnwell 1949, p. 298).

As a child Ampere is said to have calculated long arithmetical sums using pebbles and biscuit crumbs before he knew numbers. His sensitive nature was severely tried when his father was executed in the French Revolution, and eleven years later when his wife died after only five years of marriage. After recovering from these tragedies, he was able gradually to continue his career. Among many other accomplishments he, as well as Jean Baptiste Biot and François Savart, whose work supplemented his, formulated the laws named after them that describe mathematically the strength and direction of the magnetism induced in the region surrounding a wire conducting an electric current.

The next major development in electromagnetism came with the discovery of the complementary, or converse, phenomenon: not only could an electric current produce a magnetic field (the concept of a magnetic field is discussed below), but a varying magnetic field could produce an electric current. This is generally termed electromagnetic induction.

Michael Faraday in England and Joseph Henry in the United States simultaneously discovered this phenomenon in 1851, but it is more generally credited to Faraday because he published first. Son of a blacksmith, Faraday had little formal education and began his career as a bookbinder. His avid interest in natural phenomena, however, came to the attention of Sir Humphrey Davy, also a contributor to the study of electromagnetism, who sponsored the beginning of Faraday's scientific career. Among the many achievements in his career, undoubtedly the most important was the discovery (known as Faraday's law) that a changing the magnetic field encircled by a loop of wire, for example, could induce an electric current in the wire, and the faster the change, the greater the current. This phenomenon has been the basis for the prodigious development of virtually all electric machinery in industry ever since.

By permission of the Niels Bohr Library, American Institute of Physics.

JAMES CLERK MAXWELL (1831-1879)

Thus, the groundwork for understanding the intimate relationship between electricity and magnetism was laid by the insightful investigations of Oersted, Ampère, Biot, Savart, Faraday, and many others, including Dominique François Jean Arago, Joseph Henry, and Sir Humphrey Davy. What was needed, however, was an integrated mathematical description of all fundamental electromagnetic phenomena, that is, a theory unifying the electric and magnetic forces.

So, while much is made these days of efforts to unify some or all of nature's four forces, the first major unification of this kind was accomplished by James Clerk Maxwell in a paper titled *A Dynamical Theory of the Electromagnetic Field,* published in 1864. In the theory he introduced the concept of the field - the electromagnetic field - which was a logical extension of Faraday's idea of "lines of force" between electric charges or surrounding magnets.

These concepts are perhaps most easily suggested by the well-known classroom experiment of sprinkling iron filings uniformly on a piece of paper laid over a simple bar magnet with a north and a south pole. The filings will arrange themselves in a symmetric pattern that can give the impression that there is a field consisting of lines of force that extend outward from each pole of the magnet as shown in Figure 3.1.

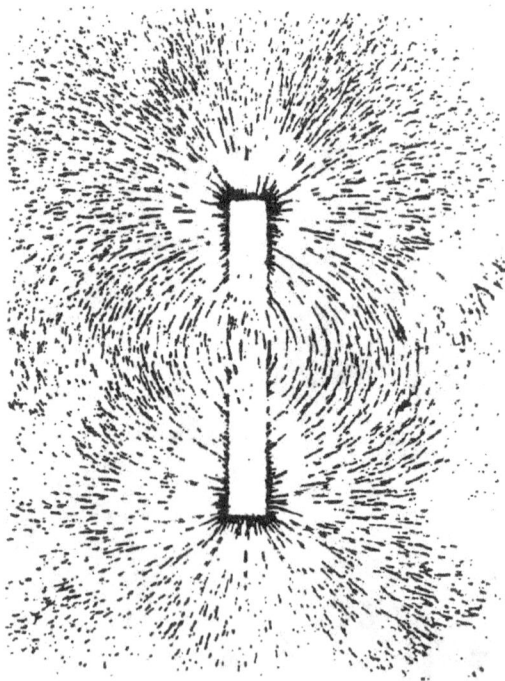

Figure 3.1 The magnetic field in the vicinity of a bar magnet as revealed by the pattern of iron filings on a sheet of paper placed over a magnet.

With a set of equations that gave mathematical expression to the field concept, Maxwell was able to give a unified description of the electric and magnetic forces. He showed that electricity and magnetism were simply intimately interactive aspects of one force: electromagnetism. His equations were later reduced by Oliver Heaviside to the four equations of elegant symmetry we know today.

These equations representing the first unification of physical forces were the beginning of an ongoing effort by physicists to seek further unification. It is well known that Einstein tried unsuccessfully to do this with electromagnetism and gravity for a major part of his life. However, in 1967 Steven Weinberg and Abdus Salam independently proposed the electroweak theory unifying the electromagnetic and weak nuclear forces. Since their work was based on a concept suggested by Sheldon Glashow, all three later shared the Nobel Prize in physics in 1979.

The electroweak theory received convincing support in 1983 from experiments at the CERN particle accelerator on the Swiss-French border and the Fermilab accelerator in Illinois. Of course it is the electromagnetic component of the electroweak theory that is of interest in this book. But the search for an experimentally verified theory incorporating the other two forces, the nuclear and gravitational, goes on, in a real sense following the tradition of unification initiated by Maxwell.

Several years ago, when I was in Edinburgh to visit a friend and colleague, I went to the James Clerk Maxwell Museum at 14 India Street, where he was born in 1831. It contains exhibits, some of them interactive, which illustrate the many contributions he made to physics, especially electromagnetism. Also shown are photographs with brief historical captions about his life.

Maxwell as a boy possessed "agile strength of limb, imperturbable courage, and profound good nature" (Richtmeyer and Kennard, 1947, p.46). He studied three years at the University of Edinburgh but finished his schooling at Cambridge University, graduating with high honors. Early on Maxwell demonstrated outstanding ability in both theoretical and experimental physics, which he used in publishing some one hundred papers on electromagnetism as well as on molecular theory and color vision. He was the first to hold the newly established professorship in experimental physics at Cambridge. There he directed the plans for the now famous Cavendish Laboratory, where many discoveries, particularly in nuclear physics, were destined to be made. His *Treatise on Electricity and Magnetism*, published in

1873, ranks with Newton's *Principia* as one of the most important works in the history of science (Richtmeyer and Kennard 1947, 46-49).

Throughout his brief forty-eight years, Maxwell was also a strongly religious man. With a Presbyterian father and Episcopalian mother, he maintained a devout Christian faith with a "strain of mysticism which has affinities with the religious traditions of the Galloway region, where he grew up" (Everitt 1975, 38). At age eight he could recite all 176 verses of Psalm 119. On recovering from a sickness while a guest at the home of Rev. C. B. Tayler, he was so deeply affected by the kind treatment he received that he saw a new meaning in the "Love of God" that left him with a lifetime conviction that "Love abideth, though knowledge vanish away" (Campbell and Garnett 1882, 170).

Maxwell composed his own daily prayers for his family, and many of his letters to his wife, Katherine Mary, were essentially prayers with frequent references to the Bible. At the end of their comprehensive biography, Lewis Campbell and William Garnett record seventy-four pages of poetry, mostly his own, several of which are religious (Campbell and Garnett 1882, 577-651). For Maxwell, his religious convictions complemented and were in harmony with his work in physics, the best-known legacy of which is the four well-known equations of classical electromagnetism that bear his name.

These four equations describe the characteristics of stationary electric and magnetic fields as well as the intimate interaction between dynamic, or changing, electric and magnetic fields. A stationary electric charge causes a stationary, unchanging electric field. When a current of such charges is moving at a constant velocity, it additionally generates a stationary magnetic field. With one exception, to be discussed below, this is the only way a stationary magnetic field can arise, because, unlike the electric charge, no magnetic charges have thus far been found to exist. So that in contrast to the electric case, a stationary magnetic field cannot be produced by a magnetic charge.

This is because all bar magnets, and magnetic materials in general, possess their magnetism by virtue of the electrons swirling about the nuclei of certain atoms such as those of iron. This motion produces a weak magnetic field, which, when aligned with its neighbors, sum to a stronger composite field.

It is as if each atom were a tiny bar magnet with a north and south pole, each atomic magnet then aligning itself with the many other atoms in the material to yield the aggregate effect of the bar magnet stuck to your refrigerator.

Because they have north and south poles, bar magnets, tiny or otherwise, are called magnetic dipoles in physics. Electric dipoles also exist, with a positive charge on one end and a negative charge on the other. The fundamental difference between the magnetic and electric cases, however, is that in the latter the positive and negative poles can be separated as positive and negative electric charges. The magnetic poles cannot be separated this way, hence, there are no magnetic charges.

However, there is an exception to the statement that magnetic fields can only exist when produced by electric currents. At the most fundamental microscopic level many elementary particles such as electrons, protons, and neutrons possess an intrinsic magnetism and behave like minuscule, microscopic bar magnets, which have north and south poles. However, as explained earlier, these poles cannot be separated and reduced to magnetic charges.

It is therefore the unit electric charge (positive on the proton and negative on the electron) that is the fundamental measure of most of the aggregate electromagnetic phenomena we experience at the macroscopic level, the level of everyday experience.

Maxwell's equations also describe how a changing electric field can generate a changing magnetic field, and vice versa, as observed experimentally by Faraday, Henry, and others.

An oscillating electric charge or an oscillating magnet can produce an electromagnetic wave of correlated and interacting electric and magnetic fields that are constantly linked and alternating, one generating the other, as depicted in Figure 3.2. One of the most important results of the theory, however, was that the radiating electromagnetic field it predicted turned out to propagate at a speed about equal to the speed of light as experimentally measured at that time. Maxwell's work not only unified electricity and magnetism, but also brought together the study of light and optics, until then considered separate disciplines, into the realm of electromagnetism.

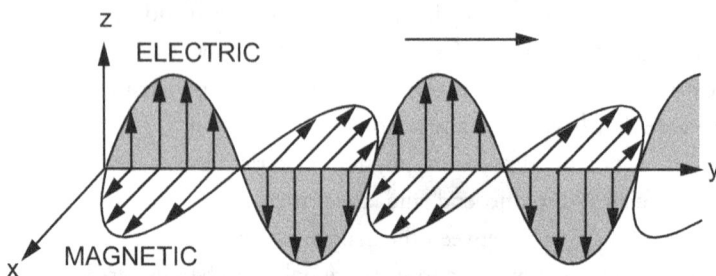

Figure 3.2 Alternating electric and magnetic fields in an electromagnetic wave moving at the speed of light along the y axis.

FORERUNNERS TO THE RELATIVITY AND THE QUANTUM THEORIES

It was not until 14 years after the publication of Maxwell's theory, however, that Heinrich Hertz produced and detected electromagnetic waves in his laboratory, thus confirming Maxwell's theoretical predictions. That same year, while studying these waves, Hertz also discovered, by accident, the photoelectric effect: the emission of electrons from a surface when it is exposed to light.

He noted that the spark between two electrodes occurred more rapidly when they were exposed to light (Richtmyer and Kennard, 1947, p. 71). Hertz was a brilliant physicist, graduating summa cum laude from the University of Berlin. But like Maxwell he died at an early age, a month before his 37th birthday. The unit of frequency, one cycle per second, is named after him.

Hertz's discovery immediately triggered the interest of a number of investigators. For example, others found that exposing a negatively charged zinc plate to light caused its charge to decrease. The search focused on determining what kind of electrically charged particle was being emitted on exposure to light. With the ingenious experiments of Philipp Lenard among others, and the discovery of the electron by J. J. Thomson, it was realized that it was an electron that the light was ejecting

Perhaps the most widely known event that marked the end of the century was the discovery of x-rays by Wilhelm Roentgen in 1895. Roentgen was studying the flow of electricity through rarefied gases in an evacuated glass tube, which was surrounded by black card board. Though the experiments were conducted in a completely darkened room, he noticed by accident that a chemically treated piece of paper became fluorescent, even two meters away (Richtmyer and Kennard, 1947, p 451-2). Using a photographic plate he and his wife saw the bones of her hand, including her wedding ring (Nobel Lectures, Physics 1901-1921).

X-rays are a higher energy form of electromagnetic radiation emitted by an atom. As a result of a collision from a high energy particle for example, an electron in one of the atom's inner quantum shells can be ejected, leaving a vacancy. The vacancy is filled by an electron from an outer shell of the atom with the emission of an x-ray. For Roentgen's discovery of this radiation, which has become an indispensable mainstay of diagnostic medicine, he was awarded the first Nobel Prize in physics in 1901

This growing scientific activity that took place in the late nineteenth century produced a wealth of ingenious experiments that were vital precursors to the quantum theory and the special theory of relativity. One of the precursors was the photoelectric effect, whose quantum nature was discovered by Einstein. Another was electromagnetic radiation, which ranged from radio waves, through infrared, visible, ultraviolet, and x-rays, to γ-rays, all of which moved at the speed of light. His insight into the nature of light, as we will see in the next chapter, made it possible for Einstein to devise the special theory of relativity.

CHAPTER FOUR

ELECTROMAGNETISM'S ROLE IN RELATIVITY AND QUANTUM THEORY

I t was a deeper understanding of the nature of electromagnetic radiation that led to Einstein's formulation of both the theory of the photoelectric effect and his theory of special relativity. At the beginning of the twentieth century the full meaning of the intense experimental work toward the end of the previous century began to be understood, paving the way for a revolution in our perception of the physical world.

LIGHT AS A PARTICLE

Einstein appreciated the fundamental implications of Max Planck's 1900 derivation of the formula for the spectrum of light emitted by the molecules of a material body at a given temperature (discussed in Chapter 1). Planck showed that the only way a mathematical expression of such a spectrum could match experimental observation was to postulate that light comes in incremental packets of energy. This was a daring departure from the century-old view that light came in the form of waves.

But Einstein also realized the importance of Lenard's ingenious experiments. They showed that increasing the frequency (number of cycles per second) of the light led to an increase in the energy of the emitted particles. Moreover they showed that increasing the intensity of the light resulted in an increase in the number of emitted particles. This direct relation between the intensity and energy of the light and the one hand and the number and energy of the electrons on the other, led to Einstein's theory of the photoelectric effect. The theory showed that light came in the form of discrete packets of energy called photons.

According to the theory, the energy of the photon is directly proportional to the frequency of the light. More precisely, the energy of a photon is equal to its frequency times what is known as Planck's constant. Planck's constant is a constant of nature; that is, it has an unchanging value. It was necessary for Planck to introduce this constant in the formulation of his spectrum in order for it to accord with experiments.

The seminal work of Planck and Einstein paved the way for the ultimate development of the quantum theory, which reached fruition in the 1920s. The theory reconciles the wave and particle theories of light. It tells us that electromagnetic radiation comes in the form of photons which, however, can statistically accumulate to collectively reveal a wave nature. This reconciliation, on hindsight, perhaps should not have been too much of a surprise, at least in a sense. This is because physicists have only been able to observe essentially two possible means of energy transport, i.e., via particles or waves (Weidner and Sells, 65, p.142-3).

THE SPEED OF LIGHT AND RELATIVITY

It is interesting to note that Einstein was awarded the Nobel Prize in physics for his work on the photoelectric effect and not for that on relativity for which he is best known. It was Einstein's insight about the nature of light that opened the way to his relativity theory. To put it in present day terms, the problem that confronted Einstein was how to show mathematically that the motion of material bodies as well as the behavior of light would be observed to be the same regardless of whether the observations were made on Earth or on a spaceship.

It was well-known experimentally since the time of Galileo that the laws of mechanics governing the motion of material bodies were the same for any observer moving at some fixed speed in a fixed direction relative to another observer with a different speed and direction. For example, suppose you throw a ball vertically in the air, it returns along the same path, and you catch it. The ball's motion is observed to be the same, and is described in the same way by the laws of mechanics whether you are standing on the ground or on a train moving at a constant velocity (constant velocity is defined as constant speed as well as constant direction). Since you, the ball, and the train are all moving together, you can still throw and catch the ball the same way.

However, while the laws describing mechanical motion (e.g., the ball's), were considered independent of the velocity of the one's frame of reference, such as the ground or a train, this was not considered to be the case for electromagnetic radiation.

This is because ever since the time of Isaac Newton, it was believed that there existed a universal medium, a fundamental background essence known as the ether which provided an absolute frame of reference. Since sound waves needed air as a medium, and ocean waves, water, it seemed logical that electromagnetic waves needed one also.

If the ether exists, however, measurements of the speed of light should vary depending on the velocity of the frame of reference in which the measurements are made. For example, a spaceship observer measuring the speed of solar light would record a different value depending on whether the ship was moving toward or away from the sun. But the true speed of light should only be observable relative to the absolute and stationary ether. So that in general it would appear to have one speed if measured on earth, which is itself moving with respect to the ether, and another if measured on a space shuttle.

This essential inconsistency of mechanical measurements being independent of the observer's velocity and light measurements not being so, brought into question the reality of the ether. Albert Michelson and Edward Morley settled the question in an experiment that compared the speed of light measured along the direction of the earth's motion to that measured perpendicular to the motion. If the ether existed, the measurements should be different. They were not. Thus nature either does not posses an ether or if it does, it is not detectable.

There has been a considerable historical controversy as to whether Einstein was aware of the Michelson-Morley experiment before formulating his theory. (Isaacson, 2007, p.116-7). Whatever the case, it was certainly true that his work was consistent with the conviction that the ether was undetectable. Accordingly, Einstein developed a theory that no longer depended on the assumption of an absolute reference frame. So that not only for measurements of mechanical motion, but also for electromagnetic radiation, all frames are on an equal footing, none are sacrosanct; all are relative. Indeed there is no way of ultimately distinguishing rest from constant velocity. In particular, all measurements of the speed of light give the same result regardless of the velocity of the frame of reference.

By thus totally democratizing the principle that all frames of reference are relative, Einstein in effect shifted the notion of absoluteness from the existence of an absolute frame of reference to an absolute maximum speed, the speed of light (Fraser 81, p.473; Harrison 85, p.144). If we think about it for a moment, it would be a strange, complex universe, if the results of a measurement of the speed of light depended on whether it was made on the moon, the earth, or a spaceship. Einstein's mathematical description of how all measurements of the speed of light are measured to be the same, no matter where you are and how fast you are moving, is called the special theory of relativity. This is because it only deals with cases where frames of reference are moving at constant velocities relative to each other. It is a special case of his general theory of relativity, formulated ten years later, which deals with reference frames accelerating with respect to each other.

In sum, through deeper consideration about the nature of electromagnetic radiation, Einstein provided vital groundwork for the development of the quantum and relativity theories. Again, he did this first by conceiving that light can exist as quanta, called photons, and second by conceiving how measurements of the speed of light can be the same in any reference frame moving at a constant velocity.

THE REFINEMENT OF ELECTROMAGNETIC THEORY: QUANTUM ELECTRODYNAMICS

The indispensable seminal work of Einstein, Planck, as well as Jules Henri Poincare and Hendrik Lorentz, and others pointed the way to the maturation of electromagnetic theory with the development of quantum electrodynamics, often simply called QED. In essence this theory reconciled Maxwell's theory for electromagnetic phenomena with the basic theory of the quantum.

Although there were many who contributed to the formulation of QED, Paul Adrien Maurice Dirac led by providing the early groundwork for the theory. He was born in 1902 in Bristol, England, the son of an English mother and a Swiss father who had moved to England from Geneva in 1888.

Partly because of his mathematical ability and partly because the upper classes of his secondary school were relatively empty due to the manpower needs of World War I, he advanced through his classes at a pace that allowed an unimpeded growth of his inherent brilliance (Kragh 1990, 1-4).

Dirac's outstanding performance won him a number of grants and scholarships that ultimately led to his studying at Cambridge, where he soon became interested in the then-new theories of relativity and quantum mechanics. Within a few years he published *The Principles of Quantum Mechanics* (1930), which to this day is considered a classic text in the field (Kragh 1990, 6ff.).

His major contribution to QED theory was the equation that bears his name and that reconciles the quantum mechanical description of the electron with the special theory of relativity. The equation employed an additional dimension over what was used in quantum theory at the time. The added dimension gave a fully symmetric description of the dynamics of an electron by predicting the existence of the positron, an electron with a positive electric charge. This particle was first observed experimentally in 1932 four years after Dirac published his work.

Using the Dirac theory, physicists could predict the excited energy states of the hydrogen atom much more accurately than they could with the earlier quantum theory that did not include relativity. I can remember when I first studied the Dirac equation how impressed I was with its simple, creative elegance. Indeed, it was a dominant feature of Dirac's philosophy of physics that "physical laws should have mathematical beauty" (Schweber 1994, 70).

It was Enrico Fermi who made the next major step in the development of QED theory. He made his own refinements of Dirac's work, which ultimately resulted in its publication in the "Reviews of Modern Physics" in 1932. It is a classic article and still studied by physicists today, and another example of Fermi's genius to express with such thoroughness and clarity his insight in virtually any branch of physics.

Building on the work of Dirac, Fermi, and others, the final formulation of the QED theory accepted today was accomplished soon after World War II by Richard Feynman, Julian Schwinger, Sin-itiro Tomonaga, and Freeman Dyson. For this work, Feynman, Schwinger, and Tomonaga were awarded the Nobel Prize in physics in 1965; unfortunately, Dyson was not included, although he made very significant contributions to the theory. All four men exhibited precocious intellects in their childhood, which were recognized and nurtured by attentive parents, and all four pursued careers characterized by prodigious study and productivity.

Born in 1918 in Far Rockaway, Long Island, Richard Feynman grew to be not only one of the most insightful and productive physicists of the twentieth century (Lubkin 1989), but also one of the most colorful. I can remember his lecture at my first attendance at a meeting of the American Physical Society in 1950. He managed to convey the physics with refreshing clarity as well as with sustained, spontaneous humor. Later in his career at the California Institute of Technology, after a long day's work, Feynman would occasionally wander into a nightclub and take over the bongo drums in the orchestra.

As a sophomore in high school he taught himself trigonometry, advanced algebra, analytic geometry, and calculus and kept voluminous notebooks. In one of his classes at MIT, the professor, because of the pressure of other duties, sometimes would not be fully prepared and would run out of notes before the end of the hour. With only a brief hesitation he would call on Feynman to complete the lecture, who did so without error and often with remarkable ingenuity (Schweber 1994, 374-75).

Feynman was gifted with incredible intuition and insight, at times arriving at a solution to a problem with just a few lines of equations where others took ten pages. Whenever he entered a conference room of colleagues and students, his charismatic blend of wit and intellectual gravitas would bring the room to life. He was a born teacher who thoroughly enjoyed sharing his knowledge, an excellent legacy of which is his three-volume lecture series. Although primarily designed for students, it can be found on the office bookshelves of most physics professors throughout the world (Feynman et al. 1963).

Julian Schwinger, born in New York City in 1918, showed an intense interest in physics at an early age. Although he took only two years to graduate from New York's prestigious Townsend Harris High School for gifted students, he created problems for his college professors because of his sporadic class attendance. Despite this Schwinger still got A's in mathematics and physics. He did not, however, do well in his other courses, preferring to spend his time pursuing his own studies. Nevertheless, his genius was recognized and encouraged, and at age sixteen he collaborated on his first paper, which was published in the *Physical Review*. In fact, he wrote his Ph.D. dissertation before he received his bachelor's degree from Columbia University in 1936 (Schweber 1994, 276-83).

Shy and introverted, Schwinger was nevertheless gentle, friendly, and spoke with remarkable clarity. I remember sitting in on one of his lectures at Harvard when I was on sabbatical at MIT. I was awed at the elegance, fluidity, and polish of his presentation, which was always the subject of admiration among his colleagues. In his early career his work routine often disconcerted his colleagues because he would sleep all day, have a breakfast of steak, french fries, and chocolate ice cream, start work at seven or eight o'clock in the evening, and leave in the morning as his colleagues were arriving (Schweber 1994, 292). Schwinger shared his prodigious knowledge and talent with two generations of physicists, and the legacy of his genius will be felt for generations to come.

Son of a well-known philosophy professor, Sin-itiro Tomonaga was born in Tokyo in 1906. As a child he was sickly and quite sensitive; nevertheless, he exhibited a strong interest in science. He and a friend made electrical circuits from scraps in a junkyard, and he boosted a 20-power microscope he was given to 200-power by a small spherical lens he made of melted glass tubing. He was a gifted teacher; his students called him "the magician." Tomonaga often said, "If you formulate the problem correctly, that is, if you ask the right question, the answer emerges spontaneously." Despite World War II and postwar deprivations of food and facilities in Japan, Tomonaga pursued his work. He translated his wartime research from Japanese into English, typing it himself on the back of used paper (Schweber 1994, 252-66).

Freeman Dyson, born in England in 1923, also revealed his precocious intellect at an early age, computing the number of atoms in the sun when he was six. Shy, but avidly interested in science, he won a scholarship to Winchester College, whose scholars are selected from the brightest, most gifted students in England. Interested in a Russian work on number theory, he learned Russian and translated it. At Cambridge his adventurism in physics calculations was occasionally complemented by engaging in the ancient tradition of "night climbing." He and a friend, Peter Sankey, would use drainpipes, chimneys, and window sills to climb many of the old buildings on campus (Schweber 1994, p. 486).

In addition to his contribution to many other branches of physics, Dyson, in the latter part of his career, expanded the horizon of his intellectual interests. He shared his insights in lectures at conferences on the relation between science and religion. Using both theories in physics and biology, he made calculations to show how it might be possible for some evolute of humankind to survive and thrive for cosmic eons as the universe continues to expand and cool (Dyson 1979, 1988).

A QED PRIMER

Each of these four men made unique contributions to the building of a complete theory of quantum electrodynamics. Approaches to the problem were based on two different viewpoints, one taking particles as the fundamental building blocks for the theory, the other taking fields - like the magnetic field around a magnet - to be the basic starting point. Feynman pursued the first approach, Tomonaga and Schwinger working independently followed the second, and Dyson showed that Feynman's results could be derived from those of Tomonaga and Schwinger (Schweber 1994, xxii-xxvi).

One of the lasting legacies of Feynman's particle approach is what are known as Feynman diagrams (Feynman 1985, 86ff.), now used by physicists all over the world to describe succinctly the elements of almost any calculation involving the behavior of interacting material particles as well as particles called bosons. Boson is the generic name for a particle that transmits the force between the material particles. The virtual photon (in the present context often simply called the photon) is the boson that transmits the electromagnetic force. Feynman diagrams find use not only with electrodynamic calculations but also with calculations involving other forces as well.

An example of a typical Feynman diagram is the electromagnetic interaction between, say, two electrons is shown in Figure 4.1. The flow of time is assumed to proceed from the bottom to the top of the page, so that the successive arrows represent the progress of each electron. Each separate line segment, straight or wavy, as well as each three-way vertex or juncture, represents a quantum mathematical expression.

For example, the solid line arrow segment at the lower left of the diagram with the arrow pointing up and to the right delineates one of the electrons as it is approaching the other electron before the interaction and is described mathematically by what is known as a wave function. Similarly, the succeeding segment pointed up and toward the left depicts the same electron, after it has interacted with its neighbor. The dashed wavy line connecting the points where the "before" and "after" electron segments meet represents the virtual photon that transmits the interaction.

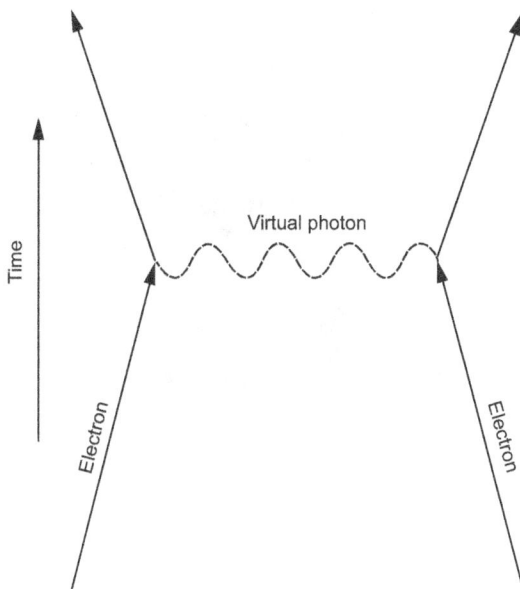

Figure 4.1 A Feynman diagram representing the mutual scattering of two electrons. Time is shown advancing in the vertical direction.

When the quantum mathematical expressions for the four straight line segments, the two junctures, and the wavy photon line are assembled in the proper order, a calculation describing the process can be made. Of course, many diagrams are much more intricate, describing quite complex interactions, and consequently involving lengthy, complicated calculations. Feynman's concept of pictorially presenting a process and at the same time specifying the prescription for assembling the mathematical components necessary to do the theoretical calculations that describe the process in detail is, however, a marvelous example of his inspiring insight.

Richard Feynman 1918-1991.
By permission of the Neils Bohr Library, American Institute of Physics. Photograph by AIP Emilio Serge Visual Archives, Weber Collection.

Tomonaga and Schwinger, in contrast to Feynman, followed an approach based on the idea that that the field is the most fundamental starting point and performed their calculations accordingly. Since for them the field is basic, they considered particles to be the quanta, or discrete localized energy materializations, of the field. For example, while the quanta of the electromagnetic field were the photons, in their approach there was also an electron field, the quanta of which were electrons.

The mathematics of both the Feynman particle method and the Schwinger-Tomonaga field theoretic approach yielded predictions in accurate agreement with experiments on electromagnetic quantum phenomena. Again, it was Dyson who showed that Feynman's theoretical results could be derived from the formulation of Schwinger and Tomonaga.

The upshot of all of the prodigious, meticulous work of Feynman, Schwinger, Tomonaga, and Dyson was a coherent system of mathematical prescriptions that could describe in exquisite detail all known electromagnetic phenomena, from the cosmic to the microscopic level. As indicated earlier, QED tells us that the electromagnetic force is carried by force-carrying photons, called virtual photons, which are actually unobservable.

This is the hidden mode of existence for photons. In contrast, the photons that help us see light and color and vivify the world around us are called real photons. While virtual photons are continually being produced, absorbed, and exchanged between bodies that interact via the electromagnetic force, real photons are generally produced in two ways.

First, if two electrically charged bodies in relative motion come close to each other, they will mutually deflect each other because of the electromagnetic force between them. For a brief time they are mutually accelerated in opposite directions. This is because the velocity of a particle involves both speed and direction; so that if either is changed, acceleration occurs. But whenever any electrically charged body is accelerated, it radiates; real photons are emitted. Thus, in the incessant motion of particles characterizing the microscopic world, such photons with a whole spectrum of frequencies are being emitted.

The second, more commonly known, means of emission for a real photon occurs when a nucleus, atom, or molecule is raised to an excited energy state, due to energy gained from a collision, for example. When the excited state decays to a lower energy state, the energy difference is emitted in the form of a real photon. The whole photonic situation is schematically depicted in Figure 4.2, which shows the virtual photon as a dashed wavy line and the real photons from the two different sources of emission as solid wavy lines.

Using the word "virtual" to specify the force-carrying photons is perhaps unfortunate because it tends to imply that they do not exist, when indeed they do. Just because we cannot observe them does not mean they are not there.

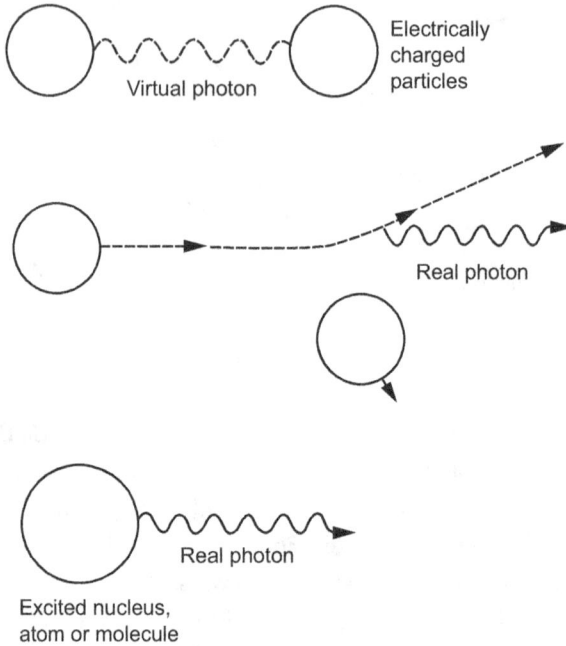

Figure 4.2 The emmission of a virtual photon (dashed wave), a real photon resulting from a collision and a real photon emitted by a nucleus, atom or molecule.

They are there and are a vital part of quantum electrodynamic calculations; and if they are not included in these calculations, we do not get the right answer - that is, the answer that agrees with experimental observations.

Another vital part of QED calculations that also involves unobservable phenomena has to do with the nature of the vacuum. If we had a perfect vacuum pump that could pump every single molecule of gas out of a sealed container, then we might reasonably suppose that the volume of the container had been reduced to a completely dormant space of nothing. But this is not true. The vacuum is alive with evanescent pairs of particles, mostly electrons and positrons that materialize for a brief instant and then vanish.

How can this be? This seems to fly in the face of the time-tested law of the conservation of energy. That is, if we use Einstein's famous formula, $E = mc^2$, we see that the masses of the electron and positron are really a form of energy and the conservation law is violated.

This, however, is where the quantum theory comes in to tell us that under certain circumstances this apparent violation is possible. At the core of the quantum theory is what is known as the Heisenberg Uncertainty Principle, which essentially states that two complementary quantities describing the state of a particle cannot simultaneously be measured with ultimate accuracy.

In other words, nature has imposed an objective limit on how precisely we see what is going on with a particle. For example, it is impossible to determine at the same time and with perfect accuracy a particle's position and momentum (momentum is mass times velocity). This means that in principle you can measure the position of a particle with great accuracy, but only at the expense of knowing very little about its momentum, and vice versa.

Another version of the uncertainty principle tells us that there is also a limit to how precisely we can simultaneously measure both the energy of a particle and the duration of time that it has that energy. This does not mean, however, that below this limit, particles that are not too massive or too long-lasting, cannot exist for a very short time, only that it is impossible for us to observe and measure them.

Thus, the law of conservation of energy applies to energy and mass we can observe, so that pairs of electrons and positrons can emerge out of the vacuum and then vanish as long as they do not stay around long enough for us to observe them. It is as if the particle pairs "borrow" energy in the form of mass from the vacuum but must pay it back quickly, and the more the mass-energy that is "borrowed," the more quickly it must be paid back (Harrison 1985, 126). In principle, more massive pairs, such as a proton-antiproton pair, can emerge, but their greater mass must be paid for by a shorter lifetime in order to remain unobservable. Thus, it is the lightest pair, the electron-positron pair that is more often present.

Nevertheless, the question immediately arises: if we can't observe them, how do we know they are there? This again is where QED comes in. For without accounting for the brief presence of such electron-positron pairs in the calculations we make to predict the results of an experiment, we do not get the right, experimentally verified answer, just as in the case of the virtual photons. The presence of both the virtual photons and the virtual particle pairs tells us, therefore, that nature, and indeed space itself, is electrically alive.

Again as discussed at end of chapter 2, we are immersed in an ocean of electromagnetic quantum undercurrents.

With the inclusion of these virtual phenomena in calculations on a host of electrodynamic phenomena, QED yields incredibly accurate answers, answers that agree with some experiments to one part in one hundred billion. Indeed, QED is by far the most accurate theory in all of physics. It is a case where humans have come as close as they may in a long time to describing accurately an aspect of nature. In Richard Feynman's words, "But so far, we have found nothing wrong with the theory of quantum electrodynamics. It is therefore, I would say, the jewel of physics - our proudest possession" (Feynman 1985, 8).

For me, this gratifying refinement of electromagnetic theory that brings it into such excellent accord with quantum theory, is but one way that I see them as related. Another way is that I cannot think of a quantum measurement being performed that does not use an instrument that involves some property of electromagnetism. It was largely the quest for knowledge about electromagnetic phenomena and their role in the atom that ultimately led to the development of the quantum theory.

CHAPTER FIVE

ELECTROMAGNETISM'S ACTIVITY IN THE MICROCOSMOS

The maturation of electromagnetic theory with the formulation of QED provides us with an exceptionally accurate description of one aspect of nature, specifically a more complete picture of what goes on in the microscopic realm. To illustrate some of the diverse activity in this realm, in this brief chapter we draw on various aspects of electromagnetic theory.

A fundamental property that distinguishes electromagnetism is that is that the particles subject to this force can have not only like electric charge, both positive or both negative, but also can opposite charge, one positive and one negative. A simple example of the latter is the hydrogen atom, which consists of a positively charged proton and a negatively charged electron.

Such charge polarity is not the case with the gravitational force. All masses attract each other; there are no positive and negative masses. This is the reason why this force is dominant cosmologically: because throughout the universe particles of opposite electric charge are constantly engaged in nullifying each other in varying degrees depending on their distance apart, even though the electromagnetic force in principle has as long a range as the gravitational. Gravitation is not so encumbered.

SUBTLE INTERACTIONS IN THE MOLECULAR UNDERWORLD

Even in electrically neutral matter, however, there can be significant electromagnetic effects that in particular make possible the binding of atoms to form a molecule. Among the simplest examples of this binding, discussed in Chapter 2, were those of common salt and of water. This kind of molecular bond, called ionic bonding, is one among several variations of such couplings basically produced by attraction between positively and negatively charged components.

Molecules in general can interact by means of still other electromagnetic effects that are weaker and more subtle. For example, a molecule, even though electrically neutral in terms of again having its number of protons balanced by an equal number of electrons whirling about the nuclei of the molecule, does not necessarily always have the shape of a perfect sphere. There can be variations in the distribution of the electrons in their orbits. These variations can result in slight concentrations of electrons in one region and depletions in another.

In the simplest case, in which there happen to be more electrons on one side than the other, the molecule can behave like an electric dipole with more negative charge on one side due to an increased concentration of electrons and a net positive charge on the other due to the an equal decrease of electrons. By virtue of this asymmetry the molecule can interact electromagnetically with other such molecules.

The electron charge distributions can even be much more complex, having, for example, a symmetrically arranged multipolar structure. These very weak, but nevertheless real, electromagnetic forces occur not only between such electrically neutral, non-spherical molecules, but also between two electrically neutral atoms or molecules that are completely spherical,

This is because during a collision the electric charge distributions of the atoms or molecules involved can be temporarily distorted into a non-spherical configuration (Weidner and Sells, 1960, p. 431). Such forces are called van der Waals forces, after the Dutch physicist, Johannes Diderik van der Waals, who discovered and studied them.

He formulated the equation describing the behavior of gases that also bears his name. It is one of the most well known equations in the kinetic theory of gases (Kennard, 1938, p. 207). For this work he received the Nobel Prize in physics in 1910.

In such molecular collisions, however, it is also true that very weak electromagnetic radiation is emitted (Amusia, 1988). Why? As discussed in Chapter 2, in any particle collision some acceleration is involved, and any time an electrically charged particle is accelerated it will emit such radiation, even an ordinarily spherical molecule that, due to the collision, acquires a distorted charge distribution. But why are particles in collision accelerated even if their speed does not change? This is because, again as noted in Chapter 4, acceleration occurs when a particles velocity is changed, and velocity consists of not only speed, but also direction.

These phenomena, however weak, revealing what happens when molecules interact in a collision, give further testimony to the incredible subtlety and range of operation of the electromagnetic force. Such subtlety also comes into play in the phenomenon of chaos.

The study of chaos and its functioning in nature has been receiving increased attention in last several decades. Chaos here is a scientific term with a different meaning than its usual colloquial one. Although any given chaotic process follows its own particular law in operation, all such processes are characterized by the fact that generally a very slight change in initial conditions can lead to drastically different results (provoking the famous comment: "A butterfly flapping its wings in Brazil can cause a tornado in Kansas") (Gleick, 1987). The minute perturbations that trigger the process to lead to one result or the other, however, all proceed via quantum electrodynamic events.

PERCEPTIONS OF ELECTROMAGNETIC RADIATION

It is unobserved evanescent virtual photons transmitting the forces that keep this page together for your reading. If we consider that indeed all of the earth's nature and a great majority of space is activated by virtual photons, and if we add to that the whole electromagnetic spectrum of real photons that allow us to see with our eyes or detect with instruments, can we posit that in a sense all of electromagnetism is essentially photons, or light. Here, as noted in Chapter 1, I again generalize light to mean all electromagnetic radiation, not just visible.

Another way to convey electromagnetic radiation's omnipresence is to look at where its radiation comes into play in various aspects of the world we experience and observe. Figure 5.1 shows a varied selection of these aspects where electromagnetic radiation operates in one way or another. Note the extensive range of frequencies as well as wavelengths that cover a range of many powers of ten. Recall that the frequency of electromagnetic radiation times its wave length equals the speed of light. For example, looking at the scale in the figure, a typical x-ray can have a wavelength of 10^{-10} meters and a frequency of 3×10^{18} cycles per second. Multiplying these two quantities together gives a speed of 3×10^8 meters per second the speed of light.

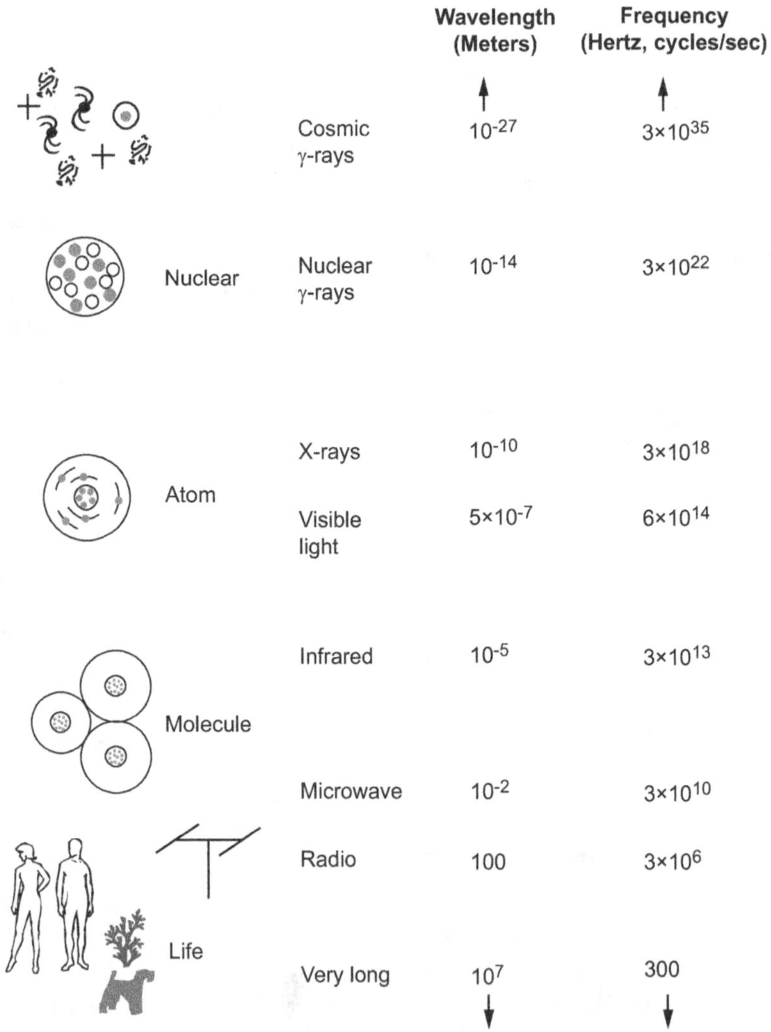

			Wavelength (Meters)	Frequency (Hertz, cycles/sec)
			↑	↑
		Cosmic γ-rays	10^{-27}	3×10^{35}
	Nuclear	Nuclear γ-rays	10^{-14}	3×10^{22}
	Atom	X-rays	10^{-10}	3×10^{18}
		Visible light	5×10^{-7}	6×10^{14}
	Molecule	Infrared	10^{-5}	3×10^{13}
		Microwave	10^{-2}	3×10^{10}
		Radio	100	3×10^6
	Life	Very long	10^7	300
			↓	↓

Figure 5.1 The electromagnetic radiation involved in different phenomena of the natural world.

FOUR PROPERTIES OF ELECTROMAGNETIC RADIATION

Another equally compelling way to perceive electromagnetic radiation's unceasing action in the micro-underworld is to examine the role played by four basic properties of electromagnetic radiation: amplitude (or intensity), wavelength (or frequency), polarization, and phase.

Intensity can vary by any tiny amount from the most subtle, animating the neural system of a fruit fly, to energizing a huge power transformer. Wavelengths can be fine-tuned with incredible precision over a virtually infinite range from the longest radio waves to the shortest gamma rays from outer space. Two waves can be in phase so that they mutually reinforce, adding to each other, or they can cancel, subtracting from each other, with all possible relative phases in between, no matter how incrementally different.

Finally, a wave can be polarized. The simplest kind of polarization, called plane polarization, occurs when a light wave oscillates in some plane perpendicular to the direction of propagation of the wave. An example is what results when light waves have filtered through your polarized sunglasses.

Another kind, circular polarization, is more complex. It occurs, for example, when light passes through a transparent material whose helical molecular structure is such that it can only transmit such light. Depending on the whether the helix is configured like a right or left handed screw, the light waves will be right or left handed circular polarized. Circular polarized light can be analyzed as a combination of two waves that are polarized in mutually perpendicular planes and 90 degrees out of phase.

Adding to the complexity, light can be elliptically polarized. All of these polarizations can be varied by an infinitesimal amount over the entire range of possible orientations. The four properties (amplitude, wavelength, phase, and polarization) in terms of light considered as a wave, and how they can vary, are illustrated in the elementary diagrams presented in Figure 5.2.

How these four fundamental properties manifesting the characteristics of electromagnetic radiation can be orchestrated in such infinite variety to provide us with such a rich panorama of nature's fecundity as well as such universal utility in modern technology is an abiding source of wonder to me. This is true from the sunlight producing the photosynthesis that greens a maple leaf to the signals coursing through our brains and nervous systems as well as the operation of our cell phones.

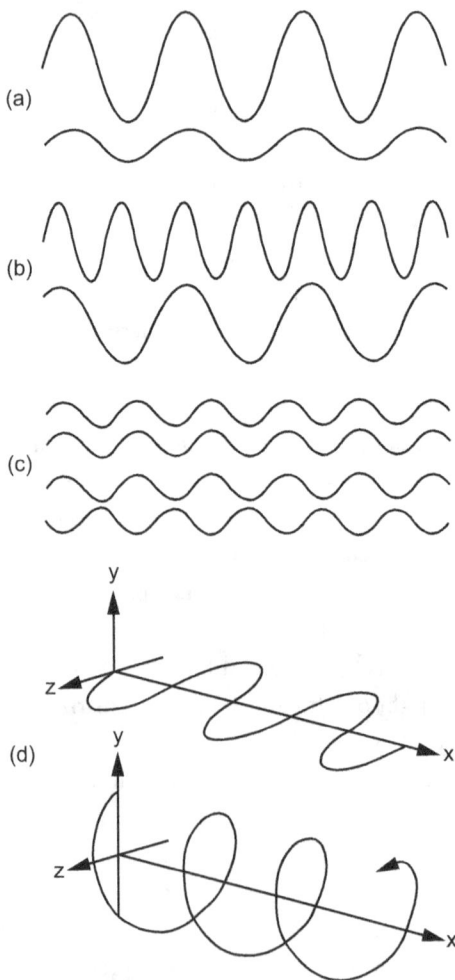

Figure 5.2 (a) Two waves of the same wavelength but different amplitudes. (b) Two waves of different wavelengths. (c) Two waves in phase and two out of phase. (d) The upper diagram represents an electromagnetic wave linear polarised in the horizontal plane. The lower diagram represents circularly polarised light moving along the x-axis. Such light can result from two waves linearly polarised in planes perpendicular to each other and 90° out of phase.

This provident variety and abundance in the activity of electromagnetic radiation on this planet is a compelling pointer to its potential for being harnessed, along with the force that generates it, for the continued burgeoning of modern technology, a technology that is fast becoming an integral part of our future evolution.

Chapter Six

Tool of Modern Technology

The number of ways that the electromagnetic force and its radiation are put to use in modern technology is legion. The best that one can do is to present a distillation of this multiplicity by selecting some representative examples that hopefully elucidate and highlight the universal utility of electromagnetic phenomena. To convey a sense of this universality I deal mostly with examples of today's technology. Nevertheless, I feel it is essential to start with technology's almost total dependence on electric power for its growth and sustenance.

Electric Power

Virtually all of modern technology makes use of electric power. Although there is a growing use of sources such as the sun and wind as well as some potential for nuclear power, most power still comes from river dams, or from the burning of bio-fuels such as coal. The burning of coal itself is an electromagnetic process. The energy latent in the complex organic molecules of coal is released in the burning. The heat is used to drive the turbines that activate electric generators which put to use Faraday's law of electromagnetic induction.

As I wrote this chapter, I found myself fantasizing how Faraday might react if he were to see the myriad of ways his law has been employed, most especially to generate electric power worldwide. Recall that Faraday found that when the magnetic field enclosed by a loop of wire is changed, a current flowed through the wire. The field can change by either varying its intensity or by varying the area of field that the loop encloses. If this area, the field's intensity, or both, are increasing, the current circulates one way; if decreasing, it flows the other way.

The basic principle of generator operation is shown in Figure 6.1. Coiled around the armature are many loops of wire, one of which is shown. The armature is caused to rotate in the field of the magnet by some mechanical means, such as a turbine. As discussed in Chapter 3, Faraday's law tells us that the current produced in such a case depends on the rate of change of the area of the magnetic field enclosed by the loop. The current is delivered by a wire to conducting brush "a" to light the lamp. When the armature rotates another 180 degrees, the wire to brush "b" is now where the wire to "a" was. Therefore the current produced is alternating (Giancoli, 05, p. 592).

At the most fundamental level this is how generators work, generators that furnish electric power all over the world. But Faraday's law of induction can be exploited for power in other, less conventional ways; for example, in alternative energy systems. An ingenious example of this, not often mentioned, is the use of the energy of ocean waves to generate electricity. The apparatus devised by engineer Annette von Jouanne is a remarkably simple.

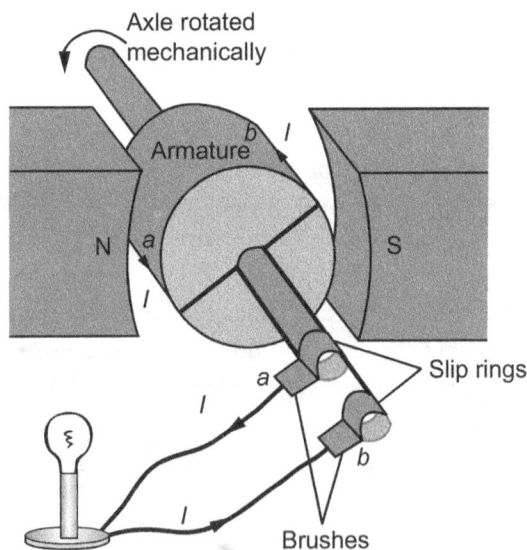

Figure 6.1 A portrayal of the principle of an alternating current (AC) generator. The electric current, I, is transmitted to the wires leading to the lamp by brushes (a) and (b) in contact with the slip rings.

©GIANCOLI, DOUGLAS C., PHYSICS: PRINCIPLES WITH APPLICATIONS VOLUME II (CH. 16-33), 6th,©2005. Printed and Electronically reproduced by permission of Pearson Education, Inc., Upper Saddle River, New Jersey.

As shown in Figure 6.2, the device is an electromagnetic buoy which consists of a tubular magnet that is attached to a floating disc. The magnet encircles a set of copper coils that are anchored to the ocean floor. As the ocean waves cause the float and magnet to move up and down, current is generated in the coils and delivered through a power line (Rusch, 09).

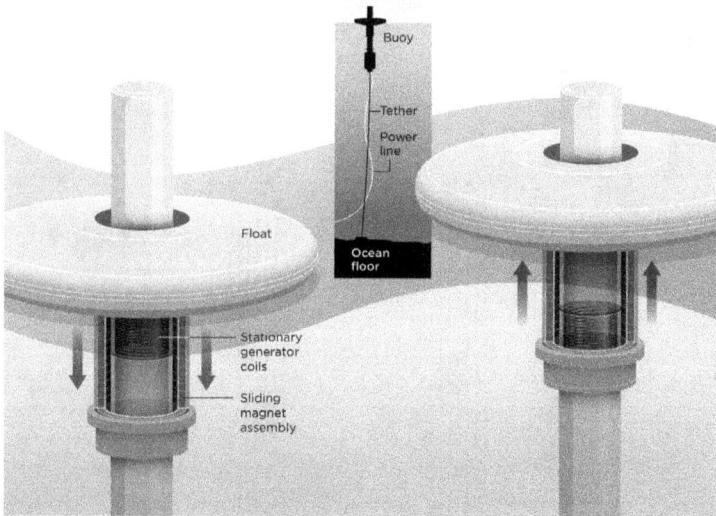

Figure 6.2 von Jouanne's wave-energy converter. Provided by 5W Infographics.

In the case of power for domestic or commercial use, the electricity produced by generators is passed to transformers that convert the power to a form suitable for delivery over long distances by high voltage power lines. Transformers are therefore an absolutely indispensable component of electric power systems worldwide.

They are necessary because if electricity were transmitted at the usual 110 or 220 volts, the large electric currents needed would result in an attendant loss in energy due to heating of the copper cables. The expense of the copper as well as that for its cooling renders such an approach completely unfeasible. Anyone feeling the cord powering an electric space heater can understand this.

Consequently, use is made of the fact that the amount of electric power is equal to the voltage times the current. So power transmitted at high voltages requires less current, and thus less copper and heat energy loss. Thus power can be transmitted over great distances to other transformers that essentially reverse the previous conversion, producing 110 or 220 volt electricity for use in factories as well as in homes for refrigerators, washers, dryers, and microwaves.

As shown in Figure 6.3, a transformer basically consists of two coils of wire, called the primary and secondary coils. They are wrapped around an iron core that concentrates, and provides a path for, the magnetic field. Such cores make it possible for nearly all (at least some 99%) of the magnetic field to pass through both coils.

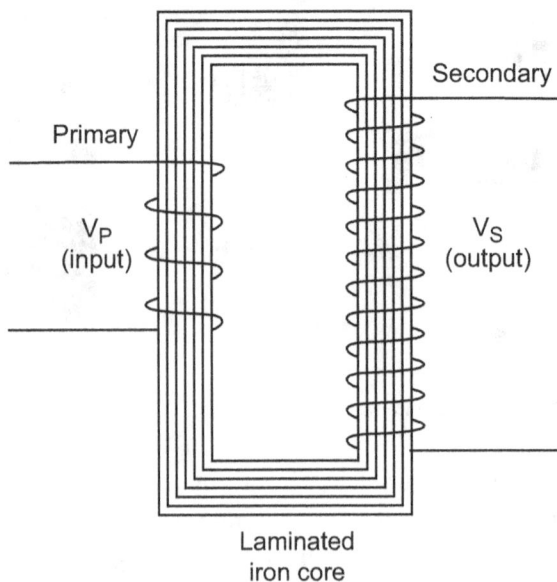

Figure 6.3 Diagram showing the principle of a transformer which steps up the primary voltage, V_p, to the secondary voltage, V_s.

An alternating (AC) voltage applied to the primary coil causes a changing magnetic field which in turn induces an alternating voltage in the secondary coil. Since the voltage of a coil depends directly on the number of turns in the coil, they can be used as step-up transformers for power transmission or step-down transformers for industrial or domestic use. Transformers are employed everywhere: TV sets, computer accessories, cell phone chargers, utility poles, etc. They put to direct use the work and discoveries of Oersted, Ampere, and Faraday.

I had my own personal experience with electric power transformers early on. Shortly after finishing undergraduate school my first job was an introduction to technology in the industrial world. I was hired by the General Electric Company as what was called a test engineer, testing electric power transformers. They were huge. The large coils of copper wire were enclosed in a steel box the size of a small room whose height, width and length dimensions were about seven or eight feet. In the top surface there was a hole almost the size of a manhole into which oil was delivered through a four-inch diameter hose.

The oil was necessary to cool the coils when they were carrying the large operating electric currents. Naturally some oil would spill on the top surface making it rather slippery, so that care was necessary with each step. One time I wasn't paying attention and the oil overflowed, spilling down the sides of the transformer onto the floor. To this day I can remember my boss standing on the floor, hands on his hips, looking up at me in utter disgust.

SILICON: THE MAINSTAY OF MODERN TECHNOLOGY

I was 21 years old then and had a lot to learn, and know that I still do. Realizing this, it was with some trepidation that, after considerable thought, I had photo-voltaic solar panels installed on a sloping berm outside of the south side of our house. Consisting of an array of specially treated silicon cells, they furnish more than three kilowatts of power for either battery storage for emergency use, or for return to the power company for credit.

The use of silicon for solar energy is but one of a plethora of applications for this vital element. In many respects, silicon does for modern technology what carbon does for living nature. Indeed, a number of philosophers of evolution speculate that we will eventually evolve into a silicon-based species much tougher than the present carbon-based one.

That silicon and carbon play such analogous roles can be understood by examining their atomic structure. Just as with carbon, silicon has four electrons and four electron vacancies or "holes" in its outermost atomic shells. This means that, just as with carbon, it can bind with two oxygen atoms to make, in this case, silicon dioxide, or sand.

In pure silicon the four outer electrons and four holes of neighboring atoms mesh to form a crystalline structure, that, at low temperatures, keeps the electrons tightly bound to their nuclei. At room temperature, however, the atoms vibrate and some electrons are freed, so that silicon becomes an electric current conductor, but not a very good one, so it is called a semiconductor. The electrons that are freed leave a hole in the crystalline structure, which in effect constitutes a positive electric charge. This is because the atom that released the electron now has a net positive charge.

When a voltage is applied to the crystal, the electrons move toward the positive terminal and the holes, acting essentially as positive electrons, move toward the negative terminal. Thus an electric current flows; the higher the temperature, the greater the number electrons released by the silicon atoms and the greater the current. Because of this temperature dependence semiconductors are used in automobile temperature gauges, thermostats, and fever thermometers.

However, most semiconductors in use today are subjected to a treatment called doping in which the silicon is deliberately contaminated, or doped, with another metallic element such as phosphorus or boron. For example, phosphorus has five electrons in its outermost shells, so that if it is substituted for a silicon atom which has four, there is an extra electron that is not bonded to an atom. When subject to this doping, silicon becomes what is called an n-type semiconductor ("n" for negative).

On the other hand, if a silicon atom is replaced by a boron atom, which has only three outer electrons, there is an electron "hole" in the crystal structure, because all of the neighboring silicon atoms have four electrons in their outer shells. This results in an effective positive charge. Silicon doped in this way is known as a p-type semiconductor ("p" for positive). Silicon treated with only a small amount of either kind of doping, becomes a much better conductor than in the case of thermal excitation in pure silicon.

The cell units in the arrays that comprise our photovoltaic solar panels contain an n-type semiconductor layer and a p-type layer. The layers are connected to power output terminals. When electromagnetic radiation from the sun strikes and penetrates the layers, electrons are released from the atoms. The voltage between one layer with more electrons and the other with more holes causes a current of these photoelectrons that yields our auxiliary power.

THE SCOPE OF THE SILICON TECHNOLOGY

I have described the specific use of silicon semiconductors for solar power in order to show in some detail with one example how they work, but admittedly I cited this example because of my belief in the use of alternative sources of power. However, probably the most essential elementary device servicing much of our electromagnetically driven technology is the silicon diode.

When the n- and p- types of semiconductors are put in contact, the electrons in the n-type will flow to the holes in the p-type, and the holes in the p-type to the electrons in the n-type until there is enough charge concentration at the juncture to resist further charge accumulation. Thus a silicon diode is formed.

As shown in Figure 6.4, diodes are one-way electronic valves. If the p-type semiconductor is connected to a positive voltage sufficient to overcome the relatively small reverse charge accumulation at the junction, a current can flow, but if connected to the n-type, there is negligible flow. Because of this property, diodes are used as current rectifiers, because they can convert a line voltage producing an alternating current to one producing the direct current required by almost all of the electronic devices in your radio, TV, calculator, and computer (Giancoli, 05, p. 830).

Light-emitting diodes, LEDs, are engineered to have, in effect, "deeper" p-type holes, so that an electron falling into a hole undergoes a greater energy loss and emits visible light. LED's are found in the displays on VCRs, CD players, clocks, and automobile instrument panels (Giancoli, 05, p. 830).

The transistor, involving the next level of sophistication in the use of semiconductors, has become the mainstay of modern electronics. Transistors can be used as switches or amplifiers and come in manifold variety. One of the simplest is the bipolar transistor that is essentially a semiconductor of one type sandwiched between two of the other. Thus there can be both npn or pnp transistors and each finds use in electronic equipment.

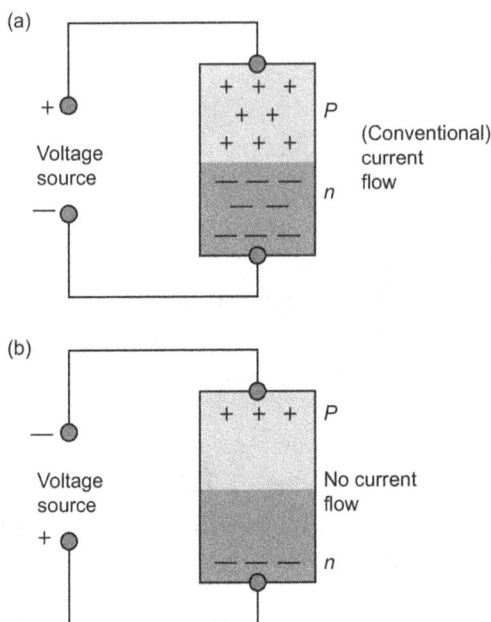

Figure 6.4 Diagram showing a p-type and an n-type silicon diode placed in contact with each other with voltage applied in two ways (a) positive voltage connected to the p-type diode resulting in conventional current flow and (b) positive voltage connected to the n-type diode resulting in no current.

©GIANCOLI, DOUGLAS C., PHYSICS: PRINCIPLES WITH APPLICATIONS VOLUME II (CH. 16-33), 6th,©2005. Printed and Electronically reproduced by permission of Pearson Education, Inc., Upper Saddle River, New Jersey.

As an example, consider the npn transistor depicted in Figure 6.5. The device has three components: the emitter (or input), the base (or gate), and the collector (or output). By applying a positive voltage signal to the p-type gate, the number of holes that are filled with electrons can be varied to regulate the current passing from the emitter to the collector. Thus a small signal at the base, or gate, can control a large current and the device serves as an amplifier. Moreover, if the gate voltage is made to operate between two extremes, a transistor can be used as a switch (Wolfson, 04, p.138 ff).

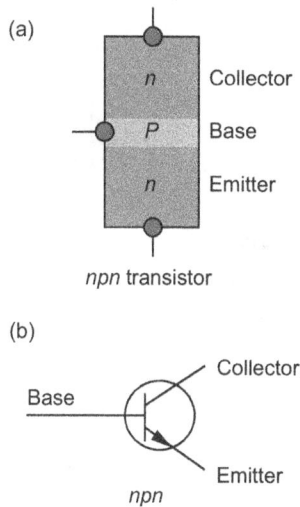

(a)

n   Collector

P   Base

n   Emitter

npn transistor

(b)

Collector

Base

Emitter

npn

Figure 6.5 Diagram of an npn transistor. A small positive voltage signal applied to the p-type base, or gate, can control a much larger current passing from the collector to the emitter so that it serves as an amplifier.

The development of transistors was a major step in the design of electronic circuits. They are much smaller than the vacuum tubes formerly used. Moreover, advances in miniaturization have revolutionized the technology so that a single transistor is now enormous compared to one that is a part of an integrated circuit, or chip. In fact a chip of the order of 1 cubic millimeter can contain thousands of transistors as well as other circuitry (Giancoli, 05, p. 831).

Only a few transistors, however, are necessary to make circuits that perform basic functions. These circuits are the elementary components which combine to make all digital electronics, and in particular, computers. Digital circuits can process and store information as binary numbers, which are sequences of bits that give a single piece of information: one or zero, yes or no, on or off, etc. (Wolfson, 04, p.175).

Such circuits are combined to make the integrated circuits that are now indispensable in calculators, computers, televisions, medical dosage monitors, and cameras. They are essential in the instruments that control automobiles, airplanes, and spacecraft. Furthermore, miniaturization makes it possible to place extremely complex circuits in a small space, and this in turn means a great increase in the speed of operation of computers, for example. This is because the distances that the electric signals must travel are so incredibly short. Thus it is myriads of tiny electric currents weaving through a lilliputian labyrinth of transistors, diodes, etc. that yield the remarkably fast response you see on your computer screen.

The relentless pursuit of electronic miniaturization, however, is approaching its limits in the work on nanotechnology in recent years. This technology generally deals with devices wherein at least one dimension is between one and one hundred nanometers. A nanometer is one billionth of a meter and an atom is of the order of tenths of a nanometer in diameter.

In addition to the developments in electronics, work is also been done on nanosensors, DNA, microscopes, and nanorobotcs, all employing electromagnetic interactions at close to atomic scales. The minute structure of material surfaces can be studied, and individual atoms manipulated, with the aid of the remarkable electromagnetic properties of laser radiation (to be discussed below). Related work has shown that copper can become transparent and aluminum combustible at nanoscale thicknesses.

However, in no other technology is electronic miniaturization exploited more universally than that of the cell phone. Cell phone transmission towers now dot much of the world's landscape conveying messages via electromagnetic radiation at microwave frequencies. The succession of ever more sophisticated versions of the cell phone, e.g. the Blackberry and I-phone, make full use of touch screen techniques. The screens are generally made of perspex, a hard plastic similar to plexiglass. When you touch the screen, your finger alters the electric field from a grid of sensors in a printed circuit on an inner layer. Successive electronic printed circuits digitally convert your message for transmission. (Woodford, 08, p.195).

The Laser

Though perhaps not quite as omnipresent as cell phones, but equal in putting electromagnetism to work, is the laser. Laser is an acronym for "light amplification by stimulated emission of radiation." Actually these words are well chosen because they state rather well what a laser is about.

The laser, which became 50 years old in 2010, exploits an atomic process, known as stimulated emission, first mathematically described by Einstein. When an atomic electron is excited to a higher energy state by a collision or by the absorption of a photon, most of the time it will immediately drop to its original ground state, giving the energy back with emission of a photon. However, many atoms have what are called metastable states. In this case the electron will remain in such a state for a considerable time before decaying to the ground state by photon emission, e.g. a millisecond instead of one hundredth of a microsecond.

The behavior of metastable states may be understood in terms of the version of the Heisenberg Uncertainty Principle as discussed in Chapter 4. Recall that it says that there is a limit to how precisely we can simultaneously measure both the energy of a state and time it remains in that state. Since the energy of a metastable state can be quite accurately measured, the principle tells us that the time it remains in that state is much less well known. Thus such states can live much longer than most other atomic states.

If the energy of a photon impinging on an atom is exactly equal to that of the atom's metastable state, the state will be stimulated to decay with the emission of a photon. So then there will be two photons that have the same energy and thus the same frequency.

Furthermore, the photons will be in phase with each other and moving in the same direction. That is, there is a resonance effect such that the second photon could not have been stimulated to emission unless it was exactly in phase and in the same direction as the first. Light with such properties is not only monochromatic (one frequency or wavelength), but also coherent (in phase).

However, for a laser to operate, the number of atoms in a confined space and in a given metastable state must exceed a certain threshold. This is accomplished by what is called optical pumping wherein light is used to raise more atoms to their metastable state than remain in their ground state.

As shown in Figure 6.6, the laser action can then proceed as a kind of chain reaction in which, for example, one photon triggers a metastable atom to release its photon. Then there are two. They stimulate two other atoms; then there are four, etc. The atoms are enclosed in a narrow cylinder with mirrors at each end. So the photons are reflected back and forth, thus maintaining the emission process. One of the mirrors, however, is partially transparent, which allows a pencil-like laser beam to be emitted.

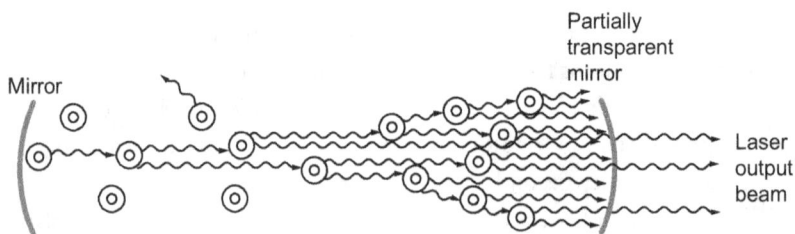

Figure 6.6 Diagram showing how radiation from atoms in a metastable state confined between two mirrors can multiply to yield a laser beam. The mirror on the right is partially transparent to allow the beam to pass through.

©GIANCOLI, DOUGLAS C., PHYSICS: PRINCIPLES WITH APPLICATIONS VOLUME II (CH. 16-33), 6th,©2005. Printed and Electronically reproduced by permission of Pearson Education, Inc., Upper Saddle River, New Jersey.

The well-known ruby laser is an example of how the atoms can be prepared in a metastable state. In a cylindrical rod of ruby, which is basically aluminum oxide, a small percentage of the aluminum atoms are replaced by chromium atoms. Intense pulses of light excite the chromium atoms to the higher of two energy states, which immediately decays to the lower metastable state as shown in Figure 6.7. Again, it is the cascade of stimulated emissions from the metastable states that deliver the thin shaft of ruby red light (Giancoli, 05, p.804).

The uses of lasers are manifold. Because of the intensity of the narrow beam, lasers are extensively utilized in surgery. Cancerous tissue in a localized region can be destroyed; the same is true for kidney and gall stones. Furthermore, a laser beam can be conducted by an optical fiber to a point where surgery is needed, such as the removal of plaque impeding the flow of blood in arteries.

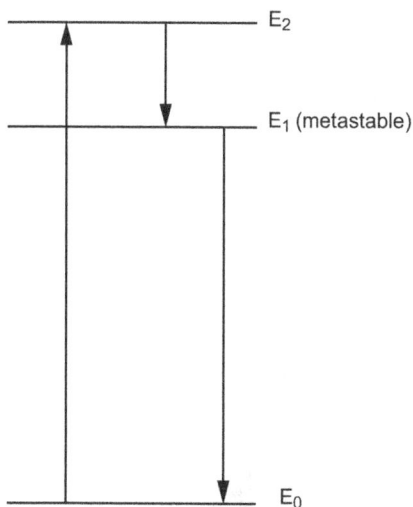

Figure 6.7 Diagram showing energy levels of the chromium atom. Photons of energy E_2 "pump" atoms from E_0 to E_2. They immediately decay to E_1 which is metastable.

Their straight-line accuracy is used for alignment in surveying as well as pointers for visual aids in the class room or for lectures generally. Because of the intense heat that can be concentrated at one spot, lasers can weld metals and drill thin holes. Every time you hear a beep at the grocery store counter a laser has read the bar code on your purchase. The laser beam is reflected from the lines and spaces of the bar code. Likewise, such a beam is reflected from the pits and spaces, representing digital 1s and 0s on your CD or DVD disc, and then decoded electronically by your audio or video system (Giancoli, 05, p. 806).

Thus lasers, needles of electromagnetic radiation, have come into universal use with the driving need for pinpoint precision in modern technology. Moreover, they are essential in our efforts to detect gravitational radiation, as seen in the next chapter dealing with how electromagnetism is used to study the other three forces of nature.

CHAPTER SEVEN

OUR EYEGLASS TO NATURE'S OTHER THREE FORCES

L aser technology is at the very core of experiments designed to detect gravitational radiation, and is a cogent example of our almost entire dependence on electromagnetic phenomena for knowledge of nature's other three forces. Although gravitational radiation is predicted in Einstein's general theory, it has never actually been detected. There is, however, indirect experimental evidence of its existence. The electromagnetic radiation from a pulsar that is orbiting another neutron star was observed for some twelve years. During this time, the time duration for one rotation of the pulsar was measured and was observed to continually decrease. Due to energy loss considered to be the result of gravitational radiation emitted by the two stars, they gradually spiraled closer to each other with increasing speed. Thus the time for one rotation was decreasing, and at a rate observed to precisely match that predicted by the general theory (Hulse 1994, p.699).

ELECTROMAGNETISM IN THE SEARCH FOR GRAVITATIONAL WAVES

For direct detection of gravity's waves, however, the precision afforded by lasers plays an indispensable role. Gravity is by far the weakest of nature's four forces, some thirty eight orders of magnitude (powers of ten) weaker than the electromagnetic force. It is only apparent and easily measurable in the case of large masses such as stars and planets. Gravity's radiation is also very weak, and is not easily observed. It thus requires extremely sensitive equipment for its detection.

There are two similar kinds of experiments utilizing lasers designed for this detection. One is ground-based called the Laser Interferometer Gravitational Observatory (LIGO), and the other is space-based called the Laser Interferometer Space Antenna (LISA). As shown in Figure 7.1, the

LIGO system is based on the same in principle as that used by Michelson and Morley briefly described in Chapter 4 to show the non-detection of ether. It essentially consists of two mutually perpendicular evacuated tubes, each four kilometers long.

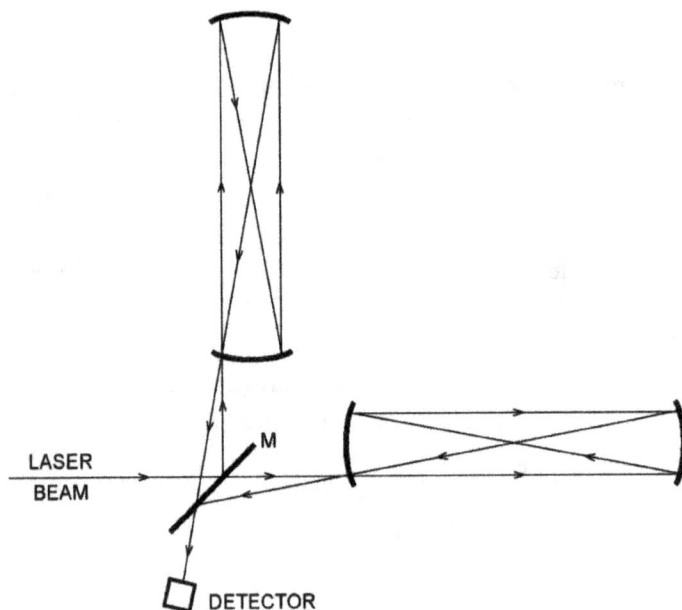

LASER
BEAM

M

DETECTOR

Figure 7.1 Diagram illustrating the basic principle of the LIGO gravitational wave detector. Laser light is split by the partially silvered mirror at M and reflected many times by the mirrors in the two arms of the system. The splitter and the mirrors are freely suspended. A gravitational wave should cause one arm to expand and the other to contract. The slightest change in the length of the arms will appear as a change in the interference pattern of the detected light.

A beam from the laser source passes through a partially silvered mirror which transmits half of the beam down one tube and reflects the other half down the other tube. In each of the tubes are specially shaped mirrors that reflect the beams back and forth many times, thus increasing the effective distance they traverse. They then pass back to the silvered mirror and thence to a photo-detector, where the interference pattern, the degree to which the wavelengths of the two beams are in or out of phase, can reveal the slightest difference in the distances they travel. All of the mirrors are freely suspended to eliminate external effects, such as environmental and seismic disturbances.

Gravitational waves are different from electromagnetic waves. They have the effect of stretching a mass in one direction and compressing it in the perpendicular direction. Such distortions cause a slight change in the length of the tubes, which in turn produces a detectable change in the interference pattern of the beams. Changes of about one hundredth of a proton radius (one hundred millionth of a nanometer) can be detected.

The LIGO project actually consists of three systems. One with 4 km arms and one with 2 km arms in Hanford, Washington, and another with 4 km arms three thousand miles away in Livingston, Louisiana. The three are operated in coincidence. This has the additional advantage of eliminating noise due to local environmental perturbations, which are unlikely to occur in all three places at once. Other such projects are GEO in Germany, VIRGO in Italy, and TAMA in Japan. Ultimately all the systems worldwide may be placed in coincidence. This affords the best chance for detection because the wave length of gravitational radiation is so long.

Although NASA withdrew its funding for LISA in April 2011, it is nevertheless an international project and due for lauch in 2015. Three synchronized laser interferometers are planned to be launched and arrayed in a triangle whose sides are five million kilometers in length. The array will follow the earth in its orbit by twenty degrees. Such a large separation of interferometers is expected to afford the detection of very long wave length gravity waves with frequencies between three thousandths and one cycle per second. LISA, therefore, should be able to detect, more easily than LIGO, gravitational radiation from such events such as the collision of two massive black holes (Miller 2010). There is even some hope that reverberations from the big bang may be observed.

The electromagnetic force and its radiation are therefore absolutely vital, and in effect are our "eyes", in the study of the force of gravity. This is true not only because of the remarkable properties of lasers, but also because all of the associated detection equipment and data collection electronics one way or another make use of electromagnetic properties. Let us see how this is so for the other two forces.

Probing the Strong Nuclear Force

I began my work in nuclear physics shortly after WWII. My PhD thesis project involved the use of a proton accelerator known as a Van de Graaf accelerator (named after its inventor) that is now at the Smithsonian Institution. It essentially consisted of a large spheroid of stainless steel mounted on a column about 20 feet high insulating it from the ground. A constantly moving, non-conducting plastic belt carried electrons away from the spheroid, leaving it with a positive charge. Inside the spheroid was a glass vessel in which hydrogen gas was ionized between positive and negative electrodes to furnish the protons that were accelerated down the evacuated tube of the accelerator.

The proton accelerator that we now see six decades later, with seven billion times the energy, is the Large Hadron Collider (LHC). A hadron is any particle with quarks as its elementary constituents and bound by the strong nuclear force. The world's highest energy particle accelerator, the LHC was constructed near Geneva in a circular tunnel 27 kilometers in circumference, straddling the French-Swiss border, and ranging from 50 to 175 meters underground.

Precisely timed and synchronized electromagnetic waves accelerate two oppositely circulating bursts of protons to within a tiny fraction of the speed of light. Some 1600 superconducting magnets keep the beams focused and on a circular path (See Figure 7.2). Most of the magnets weigh over 27 tons and are cooled by liquid helium supplied by the largest cryogenic facility in the world. At four different points, magnets alter the course of the two beams so that they collide.

In these collisions the LHC is designed to reproduce the conditions of the early universe when it was some one ten billionth of a second old. This is awesome testimony of how masterfully the electromagnetic force has been put to use. It is at the four locations where the beams collide that experiments are conducted to identify and study the properties of the elementary particles that result from the collisions.

Perhaps the most important experiment will be that involving the search for the Higgs boson which is thought to endow all particles with their mass.

Its observation would give final confirmation to the electroweak theory discussed in Chapter 3. The particle was popularized by Nobel Laureate Leon Lederman who labeled it the God particle. But implicit in the mission of the LHC is the understanding that just as much, if not more, new physics may be learned if the Higgs is not found as would be if it is. Thus the particle search will proceed with unrelenting ardor.

Figure 7.2 A view of a portion of the LHC tunnel showing the vacuum tube in which the protons are accelerated. Courtesy of Prof. Andrew Baden. Photo provided by CERN.

At the four experimental stations where the proton collisions occur, the particles in the multitude emerging from the collision events are studied. Giant detectors determine the direction of motion, mass, energy, electric charge, etc. of each particle. The detectors must be as large as they are because of the high energy of the particles. The detectors produce images of the tracks of the particles, and they are in such number that the collisions appear like a fireworks display (See Figure 7.3). Regardless of what particles are found, the utilization of the varied characteristics of electromagnetism force in their detection and identification is absolutely essential.

This is because a particle can only be directly detected if it has an electric charge. With this charge the particle can be detected by its interaction with the atoms of a detector in three ways. If (using $E=mc^2$) its energy is greater

75

than two electron masses, it can release energy by producing electron-positron pairs. Otherwise it can lose energy by collisions or by ionizing atoms. This is true whether the medium of the detector is gaseous, liquid, or solid. The charged particle will, accordingly, leave a footprint trail of electrons and ionized atoms. The detection process then involves the observation of this trail.

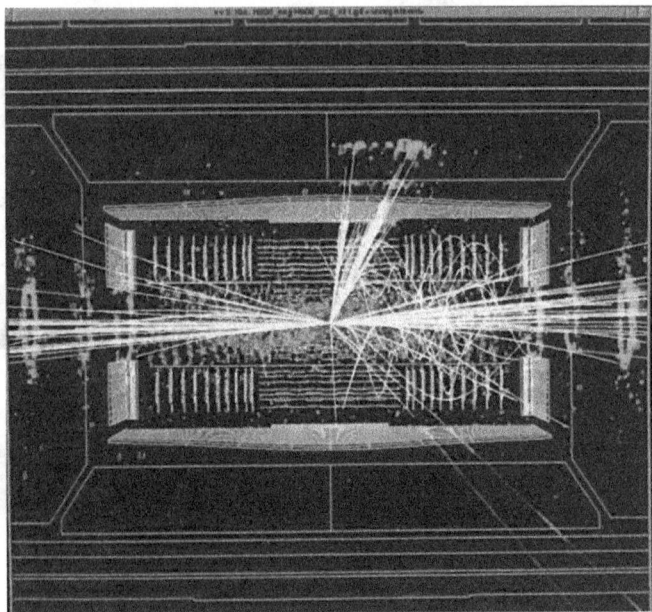

Figure 7.3 A photograph of a collision event produced at the LHC. Courtesy of Prof. Andrew Baden. Photo provided by CERN.

I can recall when I was working on my PhD thesis experiment almost 60 years ago. I used one of the most primitive particle detectors. I was detecting protons using very thick (0.4 mm) photographic emulsions. After the emulsions were developed, I would spent endless hours with a microscope recording the number and direction of the tracks made by electrons ejected from atoms ionised by the protons.

Other detectors of that era were essentially cylinders filled with a noble gas, such as argon or neon. A voltage was applied to a wire along the axis of the cylinder. Due to this voltage, a charged particle could trigger a cascade of ionization that could be electrically recorded. This is the basic description of Geiger counters which are still in use to today to detect radiation.

Although a particle's energy can often be determined by measuring the length of the track it makes, a generally more accurate measure of the energy can be achieved by passing the charged particle through a magnetic field. The field bends the trajectory of a charged particle, the amount of bending depending on the particle's energy. The higher the energy, the less the bending. Thus the energies of a stream of particles of different energies can be sorted out by the field. Moreover, the polarity of the particle's electric charge can also be determined because the field bends its trajectory, right or left, depending on its polarity, positive or negative.

The magnetic spectrometers that sort the particle energies can be huge, the higher the energy, the larger the spectrometer. Over a period of many years, I collaborated with Dutch colleagues in experiments using their linear electron accelerator in Amsterdam. It was about a tenth of a mile long with 500 times the energy of the accelerator I used for my thesis. The magnetic spectrometer there was over two stories high.

Figure 7.4 A view of the ATLAS detector at the LHC. Courtesy of Prof. Andrew Baden. Photo provided by CERN.

But even this is dwarfed, for example, by the ATLAS detector system at the LHC, which is almost 150 feet long and over 82 feet high (see Figure 7.4). It is half as big as the Notre Dame Cathedral and weighs as much as the Eiffel Tower. Designed for the Higgs particle search, it records data at the rate equivalent to 27 CD's per minute. It is an extremely complex detector system with a variety of techniques that, in one way or another, in essence observe the trails left by particles in their response to magnetic and electric fields. The ATLAS detector is a consummate example of how a full range of electromagnetic phenomena can be exploited to probe more deeply the yet unknowns of physical nature.

LOOKING AT THE WEAK NUCLEAR FORCE

Although, as discussed in Chapter 4, the electroweak theory gives a unified mathematical description of the weak and electromagnetic forces, in the work-a-day world they appear as quite distinguishable. This distinction is in part due to the neutrino, which can be regarded as a kind of signature particle for the weak nuclear force. The neutrino has a very small mass and no electric charge. It is, therefore, very difficult to detect. For this reason it was not experimentally observed until 1956, even though its existence was proposed by Wolfgang Pauli in 1930.

He proposed the neutrino to explain a puzzling feature of radioactivity. The energies of the electrons emitted in a radioactive decay did not equal the energy difference between the energy states of the nuclei before and after the decay. That is, the electron energies varied from essentially zero up to the full energy difference. Why weren't all of the energies equal to the energy difference between the nuclear states? Something was wrong. Pauli suggested that the problem could be solved if there existed a then undetectable particle, the neutrino (named by Enrico Fermi meaning "the little neutral one"), that could be emitted simultaneously with the electron and account for the missing energy.

So it is now known that in the radioactive decay of a nucleus, a neutrino is emitted along with an electron or a positron due to the weak nuclear force.

If an electron is emitted with the neutrino, a neutron in the nucleus changes to a proton, and the nucleus gains a positive charge. For example, radioactive boron with seven neutrons and five protons is transformed to stable carbon with six neutrons and six protons. On the other hand, if a positron is emitted with the neutrino, a proton changes to a neutron and the nucleus loses a positive charge.

The neutrino, with no electric charge and such a miniscule mass, is so small and weakly interactive with other matter that it can penetrate the entire earth without a collision or interaction. In one second, millions pass through your body harmlessly without your knowledge. Because of this characteristic, the chances of its detection can only be increased by looking for collision events in very large quantities of matter. That is why some neutrino detection facilities use enormous containers of water, for example, to look for the rare neutrino collisions. The containers are generally constructed deep underground in an abandoned mine in order to shield the facility from cosmic rays and other radiations that are constantly bombarding the earth that would mask the detection of a neutrino.

The neutrinos reach us from various sources such as nuclear reactions in the sun and other stars as well as supernova explosions. A neutrino traversing such a water container, on rare occasions can strike a neutron in an oxygen nucleus, for example, and cause it to change to a proton with the emission of an electron. The electron is generally so energetic that its speed exceeds that of light in water (which is measurably less than that in vacuum). This produces a cone of light in the forward direction known as Cerenkov radiation (named after its discoverer). The effect is analogous to the shock wave produced by a supersonic airplane breaking the sound barrier.

It is this Cerenkov radiation that is the signature for the passage of a neutrino. The radiation is detected through the painstaking and ingenious use of a very large number of photomultiplier tubes. The invention of the photomultiplier tube brought about a major change in experimental physics in the early 50's. As its name implies, a photomultiplier tube can multiply the effect of a photon into a recordable electric signal. As depicted in Figure 7.5, when a photon strikes the photosensitive surface, the emitted photoelectron is attracted to a succession of electrodes. Since one impinging electron can knock out a number of electrons from an electrode, the process cascades, resulting in a registered electric pulse.

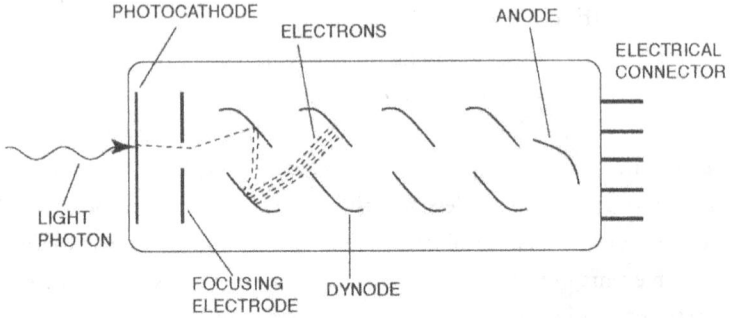

Figure 7.5 A schematic drawing of a photomultiplier. When a proton impinges on the photocathode, a photoelectron is emitted, which is focused onto the first dynode causing more electrons to be emitted. This starts a cascade that ultimately accumulates to a detectable electric pulse at the anode, which has a positive voltage.

In the underground neutrino detection facilities, the entire inner surface of the water containers is covered with an impressive array of photomultipliers in order to observe any light resulting from a neutrino collision. Figure 7.6 shows the array photomultipliers in the Super Kamiokande facility under Mount Ikenoyama west of Tokyo, which uses some 50 kilotons of ultra pure water. In Ontario, Canada, the Sudbury Neutrino Observatory, 2 kilometers deep in a nickel mine uses 1,000 metric tons of heavy water as its detecting medium (I. Semeniuk, Ski & Telescope, Sept. 2004). In heavy water a deuteron, consisting of a proton and a neutron, replaces the proton in the hydrogen atom of the water molecule. Other neutrino facilities in Italy and Russia use different detecting mediums, e.g. liquids that scintillate with the passage of a charged particle (M. Nakahata, Science, Vol. 289, 2000, p.1155).

These various experimental approaches are all devoted to continuing study of various aspects of neutrino physics. This is in part because, as far as is known today, there are three types of neutrinos. One corresponds to the electron, another, to the muon, which is an electron in an excited, short-

Figure 7.6 A photograph of a portion of the Super-Kamiokande neutrino detector array deep underground and consisting of 5,183 photomultiplier tubes. In operation the entire cavity is filled with purified water. Technicians in an inflatable raft are inspecting each of the tubes. With permission from Kamioka Observatory, Kamioka University, ICRR (Institute for Cosmic Ray Research), The University of Tokyo.

lived energy state, and the third, to the tauon, an electron in a still higher, and shorter-lived, excited state. The electron, muon, and tauon neutrinos all have the property that they can morph into one another through a process called neutrino oscillation. Because of this complex behavior, their very small mass, and their very weak interaction with matter, these three particles remain the subject of intense study as part of a new field of neutrino astronomy.

But such astronomy depends on electromagnetic phenomena for its success, from the Cerenkov radiation in water (or such radiation from scintillating fluid), to the photo-emission of the electrons in the photomultiplier, to the cascade of electrons attracted by the voltage of the

81

multiplier electrodes, to all of the electronics and computers that record the neutrino events. So regardless of which of the other three forces of nature that is investigated, it is electromagnetism that provides access to knowledge of them.

CHAPTER EIGHT

TIMING

In one way or another, imbedded in all of the projects designed to learn about the other three forces of nature, is an indispensable requirement for time synchrony and measurement for which a full range of electromagnetic properties are harnessed. This can be seen in the synchronization of the interferometer signals in the LISA project to detect gravitational waves. It is necessary for the timing of the radio-frequency waves that accelerate the protons in the LHC, as well as for the complex detecting and data taking in studying the strong nuclear force at the ATLAS facility. The large array of photomultiplier tubes in the Kamiokande neutrino detector are put in time coincidence in order to yield a coherent electric pulse signaling the passage of a neutrino.

ELECTROMAGNETIC CLOCKS

In many of cases where precise synchronization is required, the speed of light, whether in vacuum or some other medium, has been the central feature. This is especially true for the Global Positioning System (GPS) now in such extensive use. The speed of light has been measured with great accuracy, and being a speed, serves as a link between space and time. So the distance traveled by light in a given time can be precisely determined.

GPS can give a three dimensional location in terms of latitude, longitude, and altitude as well as current time. The GPS receiver calculates its position by precisely timing the signals sent by GPS satellites. At any one time there are at least 24 satellites operated by the US Government, which allows some of them for private use. Each satellite continually transmits the time of transmission and precise orbital information.

Since, in principle, three dimensions are involved, it might seem that information from three satellites would suffice for a determination of a location. However, this is not enough. A very small error in clock timing when multiplied by the very large speed of light can cause a large error in spatial location. This is why four or more satellites are used in order to apply a correction to the timing.

However, not only the clocks used in GPS, but also many other space and ground based projects, must be as accurate as possible. This is why the use of electromagnetic radiation from atoms for precision timing has been under continuous development and use for many decades.

The now famous cesium atomic clock, for example, is based on the fact that the cesium atom emits electromagnetic radiation with a frequency of 9,192,631,770 cycles per second. In fact the second is now defined as the time duration for that number of cycles of radiation to be emitted. This has become the time standard accepted universally throughout the world.

In the United States the primary standard cesium clock is located at the National Institute of Standards and Technology in Boulder, Colorado. Each time the cesium atoms accumulate the 9,192,631,770 oscillations, a beep is transmitted for the precision timing of the nation. The current version of the clock, NIST-F1, now keeps time accurate to within one second in about 100 million years.

A possible future competitor for the cesium standard is the ytterbium atomic clock under active study at NIST. It is based on some 30,000 ytterbium metal atoms cryogenically cooled to a temperature of 15 microkelvins (Bush 2009). Another contender is the so-called quantum-logic clock, which is considered to keep time to within one second every 3.7 billion years.

The precision of such clocks depends on how well they are physically isolated. The less the clock is subject to external perturbations, electromagnetic or gravitational, the more accurate it will be. Under appropriate conditions, two similar clocks experiencing the same perturbations can jointly yield more accurate time when the ratio of their frequencies (instead of their individual frequencies) is used as a standard. That is, if a disturbance affects both clocks in essentially the same way, then, although the frequency of each clock individually may be less accurate, the ratio of the two frequencies can be quite precise (Coe 69). It is an extension of this technique that is used in a global network of cesium clocks. The timing data from several hundred cesium clocks around the world are correlated to yield a standard time.

Although such exquisite precision is not required, the clocks used in many medical treatments are absolutely vital. For example, they set the rate for intravenous feeding in accord with the patient's current metabolism. They are utilized for respiratory gating in radiation treatments for lung cancer. In such gating a pulse of radiation (x-rays, gamma rays, or protons) is synchronized with the patient's breathing cycle so that the radiation strikes the same spot in the lung with each breath.

ELECTROMAGNETISM IN THE PACE OF LIFE

So the time relevance of electromagnetic interactions comes in many guises. It is such interactions that are the physical basis for the timing mechanism in the brain of an arctic tern, which makes it possible for it to navigate the round trip migration of 22,000 kilometers between the arctic and the antarctic. Electromagnetic interactions underlie the timing operations in the suprachiasmatic nuclei (SCN) in the hypothalamus that entrains our approximate living cycle. Indeed, each human person is a symphony of cosmic proportions, a symphony of information-bearing neurons, coursing blood corpuscles, all synchronized to provide the material home for consciousness, and for consciousness of being conscious.

It is this consciousness that allows our awareness of the nature of the human moment. If at a certain moment, while you are relishing your morning coffee, you also happen to be looking at the kitchen clock, you may sense that the moment you are experiencing has some duration. The second hand moved some while the moment came to full awareness. It takes some time for all of the electric signals --visual, auditory, tactile, olfactory-- to be processed by the brain and synthesized for the consciousness of an experienced moment, a moment with a diffuse, inexact beginning and ending.

Depending on the psychological test being used, the human moment has been measured to vary from about one twentieth of a second to about a full second (Whitrow 80, Fraisse 69, Macar 80). It may be apparent that this duration depends on the extent of the neural network. The time for the electromagnetic transmissions to travel through the network of a squirrel is much shorter than that for a human, which in turn is shorter than that for an elephant. One has only to observe the speed at which squirrels and elephants function to see that this is so. This is why swatting a fly is not necessarily a trivial pursuit. One way to visualize a human moment's duration is to consider that in returning the ball in a tennis match, the whole sweep of the racquet in your stroke constitutes a human moment, not just the instant the racquet strikes the ball.

So, even though the transmission speeds are very fast (but less than the speed of light), it still requires time to activate a neural network. But consider what might happen if the speed of light were infinite and our moments approached mathematically infinitesimal instants. There might no longer be the marvelous melding of the memory of the immediate past, the present, and the anticipation of the immediate future that characterizes a moment. We might not be able to enjoy a movie or integrate the successive positions of a football into a smooth trajectory. Nor could we appreciate the haunting, soothing measures of Mozart's clarinet concerto because we could not assemble the group of notes, say in a measure, and experience their synthesis as part of a living moment. We might not be able to exercise the faith, conscious or unconscious, that the moment we are living now would be followed by the next.

There also seem to be collective moments that are realized in living nature. A flock of birds feeding in a field will all suddenly take flight at a given moment. I once befriended a dog named Summer that belonged to neighbors. I asked whether I could occasionally take her with me as company while I worked on the farm. She was a cross between a collie and a sheltie and, true to her instincts, she would try to herd the cattle much to their annoyance. Though I tried to stop her, she persisted. Finally, one time when she was trying to herd, I happened to see the cattle line up in a row. Without any audible signal at a precise moment they took off as a unit in a gallop toward her; "the charge of the cattle brigade". I never saw her try to herd again.

However, for me one of the most striking examples of collective timing in the animal kingdom takes place in Maharashtra state, India, every year in late May and June. About 220 kilometers from Mumbai in the village of Purushwadi along the Kukurdi River, hosts of fireflies emerge in the evening and grace the river with a symphony of tiny lights. It is a symphony because the lights are not random, they are synchronized. For a whole stretch of the river they light up and make a bleeping sound simultaneously at a rate of roughly once a second (Lobo 2010). The phenomenon also occurs along river banks in the Malaysian jungles as well as near Elkmont, Tennessee.

The fireflies use bioluminescence, emitting light generally in the yellow green portion of the visible spectrum. In the process there is release of pheromones (sexual signaling molecules) for communication and mating selection. Giving full rein to my imagination, I see this tiny beetle with this capacity to signal, and synchronize, with its neighbors, and to emit light, as a kind of focused encapsulation of the wonder of electromagnetism and its radiation in living nature.

CHAPTER NINE

CONTEMPLATION

For a long time I have been struck by the extent of our cosmic solitude. At the end of Chapter 1 I described how, after over 50 years of searching, we have found no evidence of extraterrestrial intelligence (ETI). I discussed how extraordinary were the set of physical circumstances making possible our presence here. Even if ETI does exist we may never know it because of the finiteness of our lives and of the speed of light. These same constraints would certainly limit our capacity for reciprocal communication even if an ETI signal were observed.

I believe that it was in part this isolation and consequent sense of human centrality that prompted astronomer Edward Harrison to say: "We are the universe thinking about itself." (Harrison 1985, p.1). This privileged situation is supported by the fact that we are in a favored intermediate position in the cosmos: we can look to the vast horizons both inward to the microscopic world (discussed in Chapter 5) and outward to the cosmic world (described in Chapter 1), each with an electromagnetic "looking glass." Endowed with such special gifts of perception, it can be said that we have turned out to be the consummate example of the axiom that the whole is greater than the sum of its parts.

The form by which we exemplify this axiom, however, is still evolving, ever driven by nature's relentless reach for complexity. Some cosmologists have speculated as to how some form of intelligent life might be propagated into the indefinite future. For example, the late Robert Jastrow, who called contemporary humans "living fossils," envisioned an ever more intimate interaction between us and the instruments of technology that are even now incorporated into our lives, e.g. computers, artificial bone and joint replacements, heart pacers, etc. (Jastrow, 1980). Some even conjecture that we may ultimately evolve into a more durable silicon-based society as carbon-based humans fade away. In any case it will be the electromagnetic force that will be drawn upon to effect such an evolution.

An equally fascinating scenario that looks even further into the cosmic future has been proposed by Freeman Dyson. He starts with the fundamental assumption that consciousness finds its grounding on structure, not on matter. For example, any similarity between the human brain and a computer lies in their structure, that is, the particular arrangement and electromagnetic interlinking of the components; it does not depend on the material of which the components are made, whether carbon-based or silicon-based. So with comprehensive thought and careful calculations Dyson envisions our evolution into a more lasting embodiment than blood, flesh, and bone. He feels that it is reasonable to conjecture that we could gradually redesign ourselves to be progressively more adaptable to the ever cooler and more rarefied environment characterizing the rapidly expanding universe currently predicted by cosmologists (Dyson, 1988).

In fact, Dyson visualizes as one possibility a kind of "sentient black cloud" that has all of the memory, thought capacity, and electrodynamic communication ability of the human brain without its material frailties and thus could adapt to ever colder temperatures. Since the rate of metabolism of energy is proportional to the square of the temperature, a cold environment could be more hospitable to complex life forms than a hot one. Life is an ordered form of matter and low temperature is conducive to order (Dyson, 1988). Furthermore, by undergoing carefully chosen hibernations, less and less electromagnetic energy would be used, thus enabling such beings to exist indefinitely into the future.

Contemplating such scenarios for distant eons and their import has led me to think on another aspect of deliberating the future that was given us by John Wheeler, distinguished by his many contributions to physics, especially general relativity theory. He felt that physics had thus far developed through two major stages: the classical stage accepted until about 1900 followed by the era of quantum physics and relativity in the twentieth century; but that the next stage would manifest a quality of meaning. That he took this seriously was apparent in a number of conversations I had with him which have occasionally prompted me to consider whether there is meaning underlyng the wonder I have felt at times when I showed in a classroom demonstration how a current flows through a loop of wire when a magnet is passed through the loop.

Faraday and Maxwell have explained how this electromagnetic behavior works, and it can be described in mathematical detail. Nevertheless, I look upon the action-at-a-distance, cause and effect, of it with an almost childlike awe. This reaction is further nourished by the remarkable symmetry between the electric and magnetic forces that is apparent in Maxwell's equations. I find myself at times imagining whether there may be some clue of latent meaning that could be extracted from such striking symmetry.

Indeed, though perhaps not as apparent, there are deeper symmetries, one of which is a symmetry with respect to space. That is, Maxwell's equations can describe an electromagnetic phenomenon and a mirror reflection of it equally well. I cannot help but speculate whether this inherent symmetry is not ultimately reflected in the symmetry of a maple leaf, or even the symmetry of mammals with two eyes, two ears, and four limbs, all symmetrically arranged. I seriously question whether I could conclusively prove this statement, but my musing about its validity persists.

Modern physical theories have enjoyed remarkable success in explaining how so much of the natural world works, but there is still much to be explained regarding such things as dark matter and dark energy as well as whether Higgs particles exist. Hopefully the enigma that these phenomena present will also be solved with a satisfactory explanation. But when I reflect on the wonder I experienced on passing the magnet through the loop of wire, I feel there is yet as much mystery inherent in the explained as there is in the unexplained.

So this at root was a core motivation for my writing this book: to foster the realization that there is still much to learn right here at home on Earth, in particular with electromagnetism, the force we think we know so well. Even though we can, using quantum electrodynamics, predict some atomic phenomena with an accuracy of one part in one hundred billion, I still have a peripheral discomfort over the fact that it takes a computer calculating an almost endless chain of ever finer approximations, using ever more complex Feynman diagrams (as described in Chapter 4) to achieve such precision. Why is this so? Any physicist will tell you that physics does not explain why physical phenomena behave the way they do, but how they behave. Nevertheless, I am from time to time still haunted by the question, "Why?"

In pondering the question, I realize that electromagnetism in a sense yokes us with physical limitations. For example, as noted at the end of Chapter 2, we are immersed, and part of, an ocean of quantum electrodynamic events. Our submersion in this ocean can be considered a limitation. We cannot swim out of this ocean to see what is really going on. I believe that this situation helps in some way inform us as to who and what we are. We are creatures that are intimately related to, and dependent on, the physical world, via the rich and variegated properties of this ocean. Yet working within the limitations imposed by this situation, there is yet potential, potential for further evolution, and horizons to be explored.

Certainly one of the most engaging horizons is what is still to be learned about the human brain, which for me most cogently embodies electromagnetism's frontier. There are many scientists who are exploring this frontier by, for example, examining how awareness correlates with the synchrony of neural firing. Perhaps the most relevant here is the work of Johnjoe McFadden who claims that "the brain generates a dynamic and information-rich field that influences neuron firing through electric field coupling and its dynamics has many of the characteristics expected for a correlate consciousness." (McFadden 2002). However, in a response paper Susan Pockett finds difficulties with McFadden's electromagnetic field theory of consciousness. Among them are whether the spatial properties of the fields are measured accurately enough, and there does not seem to be a constant "relationship between brain-generated electromagnetic fields and sensation." (Pockett 2002)

I am in no position to judge the validity of this work. While it does show how the limits of electromagnetism's role in the brain are being explored, for now, to put it in somewhat philosophic terms, I find I will have to be satisfied with the belief that, for the function of the brain, electromagnetism is necessary, but may not be sufficient.

I cannot help but associate this lack of sufficiency with the reality that as accurate as the description of electromagnetic phenomena is using QED, it still is only a "map," albeit a very fine map, but it is not the terrain, terrain that we may never know completely. In the quest for an ever finer map, physicists have often been guided by response to a sense of beauty. I find this understandable since many famous physicists in their reflective writings on the revolutionary discoveries in physics in the early twentieth century have found meaning, for example, in the beauty of the mathematical formulations of physical phenomena. There is beauty to be seen in the symmetry of the electric and magnetic fields in Maxwell's equations. As discussed in Chapter 4, there is beauty in the imaginative use of an additional dimension in Dirac's equation that fully symmetrizes the description of the behavior of the electron.

Our recognition and response to beauty is deeply instinctual. We know what it is when we see it; so that beauty inherently possesses its own self-authentication. There is no question of this authentication when I go outside in April and look at the deep pink blooms on the peach and almond trees as well as the blanketing whiteness of the plum trees. Each spring these trees are a source of joy and wonder for me. While I know intellectually that it is electromagnetic activity that underlies the presence of the blooms and all of the surrounding nature and virtually all of the technology that keeps our world churning on, it is my wonder that abides.

BIBLIOGRAPHY

INTRODUCTION

Gale, Theophilus. (1628-1678), published 1802. "Electricity, or Ethereal Fire", Moffitt & Lyon, Troy, NY.

Reichenbach, Karl, Freiherr von. (1850). "Researches on Magnetism, Electricity, Heat, Light, Crystallization, and Chemical Attraction in Their Relations to the Vital Force", Taylor, Walton & Maberly, London.

Reichenbach, Karl, Freiherr von. (1851). "Physico-physiological Researches", H. Bailliere, London.

CHAPTER 1

Cowen, R. (2010). "Relic Radiation Refines Age of Cosmos" *Science News*, Feb.27, p.7.

Eisenstein D. and C. Bennett. (2008). "Cosmic Sound Waves Rule" In: *Physics Today*, Vol. 61, no. 4, April, p. 44.

Guth, A. (1997). "The Inflationary Universe", Addison Wesley, New York.

Hawking S. (1988). "A Brief History of Time", Bantam, New York.

Krauss, L. (2007). private communication.

Layser D. (1984). "Constructing the Universe", *Scientific American Library*, New York.

Lederman L. M. and D. N. Schramm. (1989). "From Quarks to Cosmos" *Scientific American Library*, New York.

Schwarzchild, B. (2006). "New Cosmic Microwave Background Results Strengthen the Case for Inflationary Big Bang Cosmology" In: *Physics Today*, Vol. 59, no. 5, May, p. 16.

CHAPTER 2

Adey, R. (1993). "Whispering between Cells: Electromagnetic Fields and Regulatory Mechanisms in Tissue" In: *Frontier Perspectives: Journal of the Center for Frontier Sciences*, Vol. 3, no. 2, p. 21-25.
Angier, N. (2009). "Sorry Vegans: Brussels Sprouts Like to live Too". Science Section, *New York Times*, December 21, 2009.
Chew, G. F. (1985). "Gentle Quantum Events as a Source of Explicate Order," In: *Zygon*, Vol. 20, p. 159-164.
Giancoli, D. (2005). *Physics*, Vol. 2 Pearson Prentice Hall, Upper Saddle River, NJ.
Hu, X. and K. Schulten. (1997). "How Nature Harvests Sunlight" In: *Physics Today*, Vol. 50, August, p. 28-34.
Kaufman S. (1995). "At Home in the Universe," Oxford University Press, New York.
Ricardo A. and J.W. Szostak. (2009). "Life on Earth" *Scientific American* September, 2009, p. 54-61.
Rue, L. (1994). "By the Grace of Guile," Oxford University Press, New York.
Shapiro, J. A. (1995). "The Smallest Cells Have Important Lessons to Teach" In: *Cosmic Beginnings and Human Ends*, eds. C. Matthews and R. Varghese. Open Court, Chicago.
Whitrow, G. J. (1980). "The Natural Philosophy of Time," Oxford University Press, NewYork.

CHAPTER 3

Campbell, L. and W. Garnett. (1882). "The Life of James Clerk Maxwell" Macmillan, London.
Everitt, C. W. (1975). "James Clerk Maxwell, Physicist and Natural Philosopher" Charles Scribner's Sons, New York.
Harnwell, G. P. (1949). "The Principles of Electricity and Magnetism" McGraw-Hill, New York.
Purcell, E. M. (1963). "Electricity and Magnetism" McGraw-Hill, New York.

Richtmeyer, F. K. and E.H. Kennard. (1947). "Introduction to Modern Physics" McGraw-Hill, New York.

CHAPTER 4

Dyson, F. J. (1979). "Time Without End: Physics and Biology in an Open Universe" In: *Reviews of Modern Physics* Vol. 51, July, p. 447-60.

Feynman, R. P., R. B. Leighton, and M. Sands. (1963). "The Feynman Lectures on Physics" Vols. 1-3, Addison-Wesley, Reading , MA.

Feynman, R. P. (1985). "Q.E.D." Princeton University Press, Princeton, NJ.

Fraser, J. T. (1981). "The Voices of Time", University of Massachusetts Press, Amherst, MA.

Harrison, E. (1985). "Masks of the Universe", Macmillan, New York.

Isaacson, W. (2007). "Einstein: His life and Universe", Simon and Schuster, New York.

Kragh, H. (1990). "Dirac: A Scientific biography", Cambridge University Press, Cambridge, U.K.

Lubkin, G. (1989). "Editorial" In: *Physics Today* Vol. 42, no. 2, p. 23.

Schweber, S. S. (1994). "Q.E.D. and the Men Who Made It", Princeton University Press, Princeton, NJ.

Weidner, R. T. and R. L. Sells. (1960). "Elementary Modern Physics", Allyn and Bacon, Boston.

CHAPTER 5

Amusia, M. Ya. (1988). "Atomic Bremsstrahlung" In: *Physics Reports* Vol. 162, p. 249-335.

Gleick, J. (1987). "Chaos", Viking, New York.

Kennard, E. H. (1938). "Kinetic Theory of Gases", McGraw-Hill, New York.

Weidner, R. T. and R. L. Sells. (1960). "Elementary Modern Physics", Allyn and Bacon, Boston.

CHAPTER 6

Giancoli, D.C. (2005). "Physics" (6th Ed.) Prentice Hall, Upper Saddle River, NJ.

Rusch, E. (2009). "Catching a Wave" In: *Smithsonian* July, p. 67.

Wolfson, R. (2004). "Physics in Your Life", The Teaching Company, Chantilly, VA.

Woodford, C. (2005). Cool Stuff and How It works", Dorling Kindersley, London.

Woodford, C. (2008). "Cool Stuff Exploded", Dorling Kindersley, London.

CHAPTER 7

Baden, A. (2011). Private communication.

Barrish, B. C. and R. Weiss. (1999). "LIGO and the Detection of Gravitational Waves" In: *Physics Today* October, p. 44.

Collins, G. P. (2008). "The Discovery Machine" In: *Scientific American* February, p. 39.

Cowen R. (2008). "Large Hadron Collider" In: *Science News* July 19, p. 17.

Feder, T. (2010). "LIGO Relocation Would Boost Gravitational-wave Science" In: *Physics Today* Vol. 63, no. 12, p. 31.

Hulse, R.A. (1994). "Discovery of the Binary Pulsar" In: *Reviews of Modern Physics* Vol. 66, p. 699-710.

Irion, R. (2000). "LIGO's Mission of Gravity" In: *Science* Vol. 288, 21 April p. 420.

Lederman L. M. and D. M. Schramm. (1989). "From Quarks to Cosmos", *Scientific American Library*, New York.

Miller, J. (2010). ""Laboratory Experiment Shows That Noise Can Be Lessened for LISA" In: *Physics Today* Vol. 63, no. 7, p. 14.

Nakahata, M. (2000). "Neutrinos Underground" In: *Science* Vol. 289, August 18, p. 1155.

Semeniuk, I. (2004). "Astronomy and the New Neutrino" In: *Sky and Telescope* September, p. 43.

CHAPTER 8

Bush, S. (2009). "NIST Improves Precision of Prototype Ytterbium Clock" In: *Electronics Weekly* August 14.

Coe, L. (1969). "The Nature of Time" In: *American Journal of Physics* Vol. 37, p. 810.

Fraise, P. (1969). "The Psychology of Time", Eyre and Spottiswoode, London.

Lobo, D. (2010). Private communication.

Macar, F. (1980). "Le Temps-Perspectives et Physiologiques", Pierre Mardagy, Brussels.

Whitrow, G. J. (1980). "The Natural Philosophy of Time", Oxford University Press, New York.

INDEX

A

accelerator
  electron 77
  proton 74
amino acids 15
Ampere, Andre Marie 26, 61
Arago, Dominique Francois Jean 27
asteroids 11
ATLAS detector 77, 78, 83
atomic shells 62
atoms 2, 14, 15, 30, 31, 41, 49, 50, 62,
    63, 66–69, 75, 76, 84

B

bacteria 17
Bell, Jocelyn 9
Big Bang Theory 1, 4
bioluminescence 87
Biot, Jean Baptiste 26, 27
black holes 9, 73
boson 42
  Higgs boson 74
butterfly 18, 21, 51

C

Campbell, Lewis 30
caterpillar 18, 19
cell phones 61, 66

Cerenkov radiation 79, 81
cesium atomic clock 84
chaos 17, 51
Chew, Geoffrey 21
chlorophyll 18
COBE satellite 4
comets 11
complexity 15–17, 53, 89
cosmic microwave background (CMB)
    3–6

D

dark energy 25, 91
Davy, Sir Humphrey 27
De Morales, Consuelo 18
Digital circuits 65
Dirac, Paul Adrien Maurice 38, 39, 93
  Dirac equation 39
DNA 16, 17, 66
Dyson, Freeman 39, 41, 42, 45, 90

E

Earth 11–12, 36, 91
Einstein, Albert 29, 33, 35–38, 46, 67,
    71
  $E=mc^2$ 75
electricity 25–27, 29, 31, 33, 58–60
  electric charge 30, 31, 39, 49, 50, 62,
    75, 77, 78, 79